LNER

K CLASS 2-6-0s

From GNR to BR

JOHN RYAN

GREAT NORTHERN

ACKNOWLEDGEMENTS

I would like to thank the following people for their help: Roger Arnold, David Burrill, John Chalcraft, Paul Chancellor, David Dippie, Hugh Parkin, Bill Reed, Andrew and Sue Warnes, David Williams.

Unless stated, all photographs from the author's collection.

Great Northern Books Limited
PO Box 1380, Bradford, BD5 5FB
www.greatnorthernbooks.co.uk

© John Ryan 2024

Every effort has been made to acknowledge correctly and contact the copyright holders of material in this book. Great Northern Books Ltd apologises for any unintentional errors or omissions, which should be notified to the publisher.

All rights reserved. No part of this book may be reproduced in any form or by any means without permission in writing from the publisher, except by a reviewer who may quote brief passages in a review.

ISBN: 978-1-914227-75-2

Design and layout: David Burrill

CIP Data
A catalogue for this book is available from the British Library

INTRODUCTION

Herbert Nigel Gresley trained as an apprentice at Crewe Works under Francis Webb, then found employment with the Lancashire & Yorkshire Railway as Assistant Works Manager at Newton Heath Carriage and Wagon Works. Whilst with the company, Gresley also received further instruction from Locomotive Superintendent J.A.F. Aspinall.

In 1904 the Carriage and Wagon Superintendent of the Great Northern Railway, E. Howlden, signalled plans for retirement and the company set about finding a successor. Despite meeting with several candidates, the GNR did not feel the applicants were suitable. Locomotive Superintendent H.A. Ivatt consulted with the Board and was able to provide the recommendation from J.A.F. Aspinall of H.N. Gresley who was subsequently offered and accepted the position. Over the next six years Ivatt and Gresley developed a good working relationship and the GNR had to look no further than Gresley when Ivatt chose to retire in 1911.

Ivatt had done much to improve and modernise the locomotive stock over his tenure and Gresley inherited a strong group of locomotives. Yet, there was a gap for a mixed traffic engine capable of working at relatively high speeds. In early 1912 the number of passenger locomotives being used on freight duties was considered far too high and Gresley was instructed to produce a suitable design for construction.

Most goods services were heretofore operated by 0-6-0 engines, though a recent development was the introduction of 2-6-0 locomotives to Britain. The wheel arrangement had been first used in America during the early 1840s and was quite popular in the country to the end of the century owing to the reduced stresses encountered at the front end. This was thanks to the pony truck helping to reduce the effect of imperfections in the track. In Britain, the Great Eastern Railway was the first company to use 2-6-0s on their system, whilst private builders also produced the type for use overseas. At the start of the 20th century, the Great Western Railway built a number of 2600 Class locomotives with inside cylinders and outside frames under William Dean, whilst his successor, G.J. Churchward, modified the design and ordered a large amount of 4300 Class engines from 1911 through to Grouping.

The GNR had experience with the type through a crisis in construction. With full orders at Doncaster and unrest in the workforce of private contractors, the company turned to the Baldwin Locomotive Works, Philadelphia, USA, to quickly produce 20 2-6-0s to be delivered at the turn of the century. These proved satisfactory, despite many American features, on goods services in the West Riding of Yorkshire, also being trialled on suburban services in London. The H1 Class 2-6-0 had a short service life and were all condemned by the mid-1910s.

Gresley did not hesitate to adopt the relatively unused wheel arrangement for his first solo design. The leading pony truck served several purposes, such as improving stability at higher speeds, reducing wear at the front of the engine, providing greater support for the smokebox and cylinders, etc. For the pony truck, Gresley developed a new suspension which he patented and used for the later P1, P2, V2 Classes, amongst others.

Carrying the weight of the front end was particularly important for the new locomotives as Gresley continued the practice of equipping the boilers with a superheater. Starting under Ivatt towards the end of his tenure, the feature improved efficiency by using the hot gasses to further heat the steam and increase expansion to obtain more work by a lower volume of steam. Unlike Churchward on the GWR, Gresley did not move towards higher working pressures for the boiler and kept with a modest 170 lb per sq. in until after the First World War. Another relatively uncommon feature for the time was the use of Walschaerts valve gear.

The first engine was ordered from Doncaster during March 1912 and was in traffic by mid-August. Classified H2, locomotive no. 1630 was dispatched to King's Cross shed for trials and was immediately found suitable for the assigned tasks. A further nine H2 engines were ordered in the same month and erected at Doncaster between February and April 1913. The ten were distributed between King's Cross, Doncaster and Colwick sheds for fast goods and secondary passenger trains.

Though the H2s were a success, further improvement was possible with the use of a larger boiler with more superheater elements. This was designed during 1913 in preparation for another order placed at Doncaster in August. The diameter was increased from 4 ft 8 in. to 5 ft 6 in. and the superheater elements used rose from 18 to 24. Some other detail changes were made for engines nos 1640 to 1649 which were completed between April and June 1914. The new locomotives were classified H3 and displaced some of the H2s from the main sheds. The H3s were also built in time to assist with the traffic associated with the First World War. Yet, the conflict vastly reduced the capacity of Doncaster Works owing to munitions and war material production. The next ten H3s ordered in July 1914 were delayed by two years before sent into traffic, whilst another 45 engines had to be purchased from contractors and these appeared in 1918 and 1921.

During the war, Gresley developed his ideas for future construction once the hostilities ended. To obtain more power for increasingly heavy trains, he chose to add a third cylinder and designed a method of driving the valve for the centre cylinder using the outside valve gear. As a trial, this arrangement was fitted to a new O2 Class 2-8-0 locomotive, no. 461, in 1918 and immediately impressed Gresley who boldly declared he would concentrate on three-cylinder locomotive designs in future.

Around this time, Gresley was also considering a new mixed traffic engine which was around a third more powerful than the H2 and H3 Classes. The design developed to use three

18½ in. diameter cylinders with a 6 ft diameter boiler – the largest used by a British locomotive at the time – and 32-element superheater. The first engine was no. 1000, classified H4, which appeared from Doncaster in March 1920 and followed by another nine examples to August 1921. The group was split between King's Cross, Peterborough and Doncaster for express goods and some important passenger duties. Experience with the H4s was valuable for the development of Gresley's A1 Class Pacific which appeared in 1922 before the end of the Great Northern Railway.

The First World War saw the British railway companies placed under Government control for the duration and when the Armistice came, thoughts turned to the reorganisation of the network to improve operations. The solution delivered was the creation of four separate companies covering the main lines of Britain. The Great Northern Railway was superseded by the London & North Eastern Railway from 1st January 1923. Of the candidates for the position of Chief Mechanical Engineer, Gresley was the most suitable and took the post later in the year.

Inheriting a diverse group of over 6,000 locomotives, Gresley had the task of moving the stock forward to meet the needs of the LNER. He was able to do this partially by adapting his own existing designs, such as the GNR H4. Two groups of 25 were ordered from Darlington Works in October and November 1923, being amongst the earliest placed by the LNER, and reclassified K3. The other Gresley 2-6-0s, H2 and H3, were also changed to K1 and K2 respectively. The new K3s conformed to the LNER loading gauge and were also provided with a cab. Design work was carried out at Darlington which drew inspiration from the North Eastern Railway's contemporary designs. The K3s had a new tender as well which later became Group Standard for several classes. The LNER K3s were distributed to the North Eastern Section and Scotland, as well as the ex-GNR main line.

The influx of a large number of K3 Class engines displaced K1s and K2s into other areas. Several went to the Great Eastern Section at March and were later permitted to run in East England. The Great Central Section had a group dispatched to Annesley and Mexborough. Over a dozen K2s went to Glasgow Eastfield for use on the West Highland Line where they became mainstays for many years. These were later provided cabs for the comfort of enginemen and also named after lochs in the surrounding area. A small group resided at Edinburgh St Margaret's shed for use on the Waverley Route, northward to Perth and Dundee, as well as to Newcastle.

In the late 1920s and 1930s, a further 123 K3 Class locomotives were built bringing the total to 193 examples at work for the LNER. Many had detail differences owing to location of construction and improvements made to the design, etc. The class tended to congregate at the main points of the sections, such as King's Cross, Peterborough New England, Doncaster, Gorton, York, Heaton and Carlisle.

Towards the end of his tenure, Gresley designed a new locomotive to work on the West Highland Line as the K2s often had to be paired with another class member or a Reid D34 Class 4-4-0 when the trains proved too heavy. The first K4 Class engine, no. 3441 *Loch Long*, appeared in 1936 and another order for five was completed in 1938.

When Gresley died in office during 1941, he was succeeded by Edward Thompson, who had a different opinion of locomotive design. He also had to react to changing servicing practices experienced in the Second World War. Thompson proposed to eliminate the use of three cylinders on all but the express passenger classes and concentrate on two-cylinder locomotives. A K3 Class engine, no. 206, was rebuilt with this arrangement and boiler with higher working pressure to compensate for the loss of a cylinder. In trials the rebuilt engine, classified K5, produced an improvement over the fuel consumption of K3s, though by this time Thompson had retired and no further action was taken.

Similarly, a K4 was reconstructed with two cylinders in late 1945 and became a prototype for a new class, the K1. When Thompson retired in 1946, A.H. Peppercorn became CME and slightly modified the design of the K4 rebuild for a production series of 70 locomotives ordered before the end of the LNER in late 1947. These engines were contracted to the North British Locomotive Company and the batch was delivered during 1949-1950. The new K1s were originally used at March shed for East England traffic and in the North East, being considered a replacement for 0-6-0s on certain duties.

Under British Railways, 344 K Class 2-6-0s were operational across the ex-LNER system. This number started to fall from the mid-1950s as the original GNR H3s (K2s) were withdrawn for scrap. British Railways' Standard Classes took over many duties and in the late 1950s dieselisation began, with the East Coast Main Line's southern section and East England amongst the first systems to be steam free in the early 1960s. The K Classes were particularly affected, with K2 and K3 classes extinct by 1961.

A nobleman enthusiast was an early pioneer in the preservation of steam locomotives and bought K4 Class no. 61994 *The Great Marquess* in 1962. The last K1 in traffic, no. 62005, was stored for a time with a view to donating the boiler to no. 61994 when needed. Thankfully this did not occur and no. 62005 was bought and restored by the North Eastern Locomotive Preservation Group. The locomotive has been prominent on the North Yorkshire Moors Railway and is currently under overhaul.

No. 61994 is presently permanently out of service in a museum leaving the LNER K Class unrepresented in the national network. Hopefully this collection serves to honour a hardworking group of locomotives overshadowed by more famous designs.

John Ryan
Over Peover, June 2024

Great Northern Railway H2, H3, H4 Classes

Above **H2 CLASS NO. 1630**
Decorated with works grey livery and standing in the yard outside the Crimpsall Repair Shop at Doncaster Works, no. 1630 is pictured when new during August 1912. The locomotive was the first to use Gresley's double swing link suspension for the pony truck which was designed to give even weight distribution as the front end changed direction. No. 1630 was unique in having 3 ft 8 in. diameter wheels for the pony truck as later examples had 3 ft 2 in. diameter. Also, early H2 Class locomotives utilised laminated springs before changing to helical springs.

Above **H2 CLASS LOCOMOTIVE**
A H2 Class locomotive's journey through the New Erecting Shop has reached the end here in late March 1913, likely being one of nos 1638-1639 which were in service during April. As built, the H2s had a boiler 11 ft 11⅞ in. long, 4 ft 9⅛ in. diameter, working at 170 lb per sq. in., though the latter was subsequently increased to 180 lb per sq. in. Also a feature when new was the position of the firebox wash-out plugs which were spaced evenly, though a staggered arrangement was present on some by Grouping.

Opposite above **H2 CLASS NO. 1630**
In the early days of the railways, the locomotive's valves were mostly operated by Stephenson valve gear which was relatively simple and easy to use. In the mid-1840s, Egide Walschaerts, a Belgian engineer, developed a new type bearing his surname that allowed greater control of valve events and this in turn improved efficiency. Railway designers mostly favoured Stephenson motion for many years and Walschaerts valve gear was unused until 1873 when a Belgian locomotive received the equipment. The first British engine fitted was a Fairlie 0-4-4T which was an exhibition piece later bought by the Swindon, Marlborough & Andover Railway. In the early 20th century, Walschaerts motion grew in popularity to become the dominant valve gear internationally. Gresley was quick to use the type and no. 1630 was the first of his locomotives fitted, though in a slightly modified form. The engine is seen here in works grey when new and the motion is highlighted in white.

Opposite below **H2 CLASS NO. 1633**
No. 1630 evidently proved successful immediately as during August 1912, a further nine locomotives were ordered from Doncaster Works. In December one of these engines, no. 1633, is in the course of erection and work looks to be concentrated on the leading axlebox. The journals were 8½ in. diameter and 9 in. long and lubrication was originally delivered by siphon oil boxes. Later, Wakefield mechanical lubricators were fitted, with no. 1633 a recipient in October 1923 as the last to be so treated. No. 1633 was complete during February 1913.

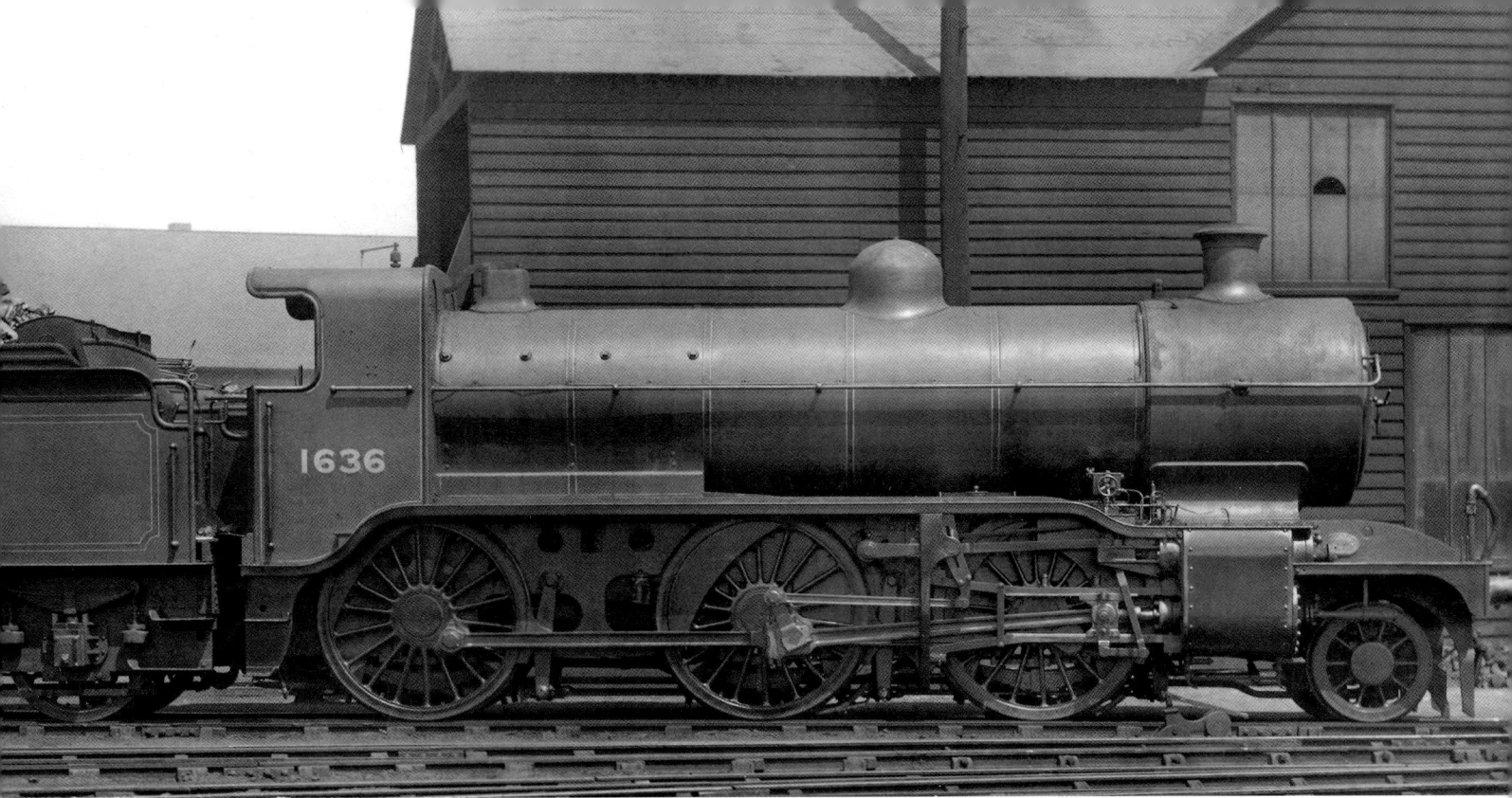

Above **H2 CLASS NO. 1636**
No. 1636 sits behind the coal stage at King's Cross shed early in the engine's career. A feature visible that changed from no. 1630 concerns the frames. Originally, just two lightening holes were present at the firebox end, then for the production batch a second pair of smaller holes were drilled to allow access to two firebox wash-out plugs.

Opposite above **H2 CLASS NO. 1634**
No. 1634 left Doncaster Works in February 1913 as works no. 1375. The H2s were initially distributed between King's Cross (3), Colwick (6) and Doncaster (1), though in 1914 the trio in London was reallocated. Movements likely occurred during the First World War yet records of these have failed to survive and no. 1634 was first recorded as being allocated to Ardsley shed before Grouping. At an unrecorded location here, the engine is seen before November 1915 when fitted with a mechanical lubricator for the axleboxes. This would be placed on the running plate over the centre coupled wheel.

Opposite below **H2 CLASS NO. 1635**
At Nottingham Victoria station with a local train in the early 1920s is no. 1635. The locomotive was built at Doncaster in March 1913 and appeared in the standard livery of the time which consisted of apple green with black and white lining whilst frames, footsteps and buffer shanks were brown/red. Under the London & North Eastern Railway the H2 Class had a standard livery of black lined with red bands. No. 1635 would soon receive this, likely when the new number – 4635 – was used from late January 1925. Photograph by T.G. Hepburn from Rail Archive Stephenson courtesy Rail-Online.

Opposite page and above **H3 CLASS NO. 1640**
Soon after the introduction of the H2 Class, the decision was made to design a larger boiler for use on a second batch of ten 2-6-0s to be built in 1914. The boiler was 5 ft 6 in. diameter and had a 24-element Robinson superheater, which replaced the Schmidt 18-element arrangement used with the first ten. There were other detail differences and the cab was slightly modified owing to the increased size of the boiler. The roof was extended rearwards 9 in. for increased protection of enginemen, whilst the forward look-out windows changed from circular to contoured, following the shape of the rear firebox. No. 1640 was the first of the group, classified H3, completed at Doncaster during April 1914 and the three images here are of the engine when new.

Above **H3 CLASS NO. 1680**
As goods traffic increased during the First World War, more engines were necessary. In the middle of the conflict, 20 H3s were contracted to the NBLC owing to capacities at Doncaster Works being reached through war-related activities. Then, a batch of 25 was ordered from Kitson & Co., Leeds, in 1919 as the pressure was still felt and these were delivered between June and September 1921. Interestingly, these cost the GNR £10,000 each, when just four years earlier, the same company, as well as the Vulcan Foundry, NBLC and Beyer, Peacock & Co., had quoted just over £4,000 for individual locomotives. Numbered 1680 to 1704, there were some detail differences, such as outside steam pipes, transverse stays in the firebox, new type of piston, cast-iron piston heads without tail rods, Skefko roller bearings on the eccentric return crank pin, Ross pop safety valves, etc. No. 1680 is seen here when new in works grey livery.

Opposite above **H3 CLASS NO. 1668**
No. 1668 was built at the North British Locomotive Company's Hyde Park Works in August 1918. Seen some three years later, the locomotive is at King's Cross shed with a non-standard feature. The economic post-war situation in Britain was poor leading to reduced consumption. Colliery owners wished to cut wages which led to a strike lasting around three months. The action reduced production, affecting the railways and some locomotives were fitted for oil firing. The storage tank sits in the tender here and possessed a 1,000-gallon capacity. Seven H2/H3 Class locomotives were equipped from around mid-1921 to the end of the year. Also of note is the tender livery which should have featured lining to the edges. Photograph by W.H. Whitworth from Rail Archive Stephenson courtesy Rail-Online.

Opposite below **H3 CLASS NO. 1655**
A group of engineers pose with no. 1655 at King's Cross shed, c. 1920. The locomotive was part of the second batch of ten H3 Class locomotives built at Doncaster in 1916, being ready for service during October 1916. A feature of these engines was the enlargement of the connecting rod big end bearing surfaces from 5 in. to 5½ in. This resulted in thicker flanges for the cylinders to keep the centre line correct. No. 1655 was unique in the group being fitted with Skefko roller bearings on the eccentric return crank pin. The feature later became standard for the last 25 H3s built. No. 1655 was noted as being at King's Cross shed from late 1920 to 1925 when moved to Doncaster.

Above **H3 CLASS NO. 1703**
No. 1703 stands outside King's Cross shed with classmate no. 1669 partially visible on the right.

Below **H3 CLASS NO. 1701**
No. 1701 was new to Peterborough New England shed in August 1921 and had employment there to 1925. Duties at the depot included fast freight trains, partially fitted goods, coal empties, as well as stopping passenger services.

Above **H4 CLASS NO. 1000**
In March 1920 the first H4 Class locomotive was completed at Doncaster Works. The locomotive's cab is featured here and a new item over the H2s and H3s was the pull-out regulator handle on each side of the firebox, whereas the earlier engines had a horizontal lever. The reversing screw column also had vacuum operated clutch gear which set the motion in position and prevented unwanted slips in the apparatus. The cab was 8 ft wide, and this was 9 in. narrower than the running plate. Both driver and fireman's sides were slightly raised from the central portion in front of the firehole door.

Above **H4 CLASS NO. 1003**
With the First World War changing regular railway operations and traffic, Gresley had time to develop design ideas as he also grew into his position. He first experimented with three cylinders in a 2-8-0 which he immediately adopted for future engines. Around this time, Gresley was thinking of a more powerful 2-6-0 and instead of increasing cylinder size, which he considered initially, three cylinders were chosen, with a larger boiler to provide ample steam. At the end of the war, work progressed on the design and an initial batch of ten was ordered from Doncaster. These locomotives, the H4 Class, appeared between March 1920 and August 1921. No. 1003 was erected during October 1920 and is seen here before Grouping at New Southgate (around six miles from King's Cross station) with an express passenger train. A local train led by N2 Class 0-6-2T no. 1728 has been held on the right with a suburban train that had called at New Southgate station, just behind the road bridge. Photograph by H. Gordon Tidey from Rail Archive Stephenson courtesy Rail-Online.

Below **H4 CLASS NO. 1001**
No. 1001 was completed about three months after the pioneer H4 and is nearly new here in the yard outside King's Cross shed. The engine stands impressively thanks to the 6 ft diameter boiler which was the largest used by a British locomotive at the time. Both no. 1000 and no. 1001 were based initially at Peterborough New England shed and rostered on Grimsby fish trains as well as goods services to London.

Above **K2 CLASS NO. 1671N**
Several trains ran daily between King's Cross and Cambridge and the 17.30 service has been caught here at Potters Bar during 1925. K2 Class no. 1671N leads the stock northward which features a quadruplet articulated set, along with some more six wheelers and bogie stock at the rear. With the change of company ownership, the locomotive has been rebranded and is now using the suffix letter owing to the existence of duplicate numbers. When this issue was rectified, the engine became no. 4671 after running for around two years as no. 1671N. Photograph by F.R. Hebron from Rail Archive Stephenson courtesy Rail-Online.

London & North Eastern Railway K1 and K2 Classes

Below **K2 CLASS NO. 1666**
No. 1666 was built by the NBLC at their Hyde Park Works in July 1918. At this time the locomotive, and others part of the order, had a grey livery with white lining but not for the tender. Standing in the yard at King's Cross shed in mid-1923, the engine had recently undergone a general repair at Doncaster which saw the new black livery with red lining applied, as well as the 'L&NER' company branding on the tender. The ampersand in the title lasted a short period from around March 1923 to June 1923 when discontinued. No. 1666 was allocated to King's Cross shed until October 1925 when moved to Doncaster depot. Photograph by W.V. Wiseman from Rail Archive Stephenson courtesy Rail-Online.

Below K1 CLASS NO. 4636

After Grouping, a certain amount of integration occurred in the locomotives fleet as useful designs were sent to areas where the native population was lacking suitable engines. In the case of the K1s and K2s, the introduction of the K3s displaced the earlier designs from the traditional route of the GNR main line to other systems such as the ex-Great Eastern, ex-Great Central, etc. Mexborough shed belonged to the latter and the first GNR 2-6-0s were sent there in May 1929 and rose to six class members, subsequently settling at five to 1940, which were utilised on local passenger services. These ranged from Cleethorpes to Hull, Penistone, Sheffield, etc. No. 4635 arrived in the second batch of two locomotives during 1930 and stands in the shed yard in September 1932. This was four years before rebuilding, though the engine has lost the piston tail rods. Photograph by C.L. Turner from Rail Archive Stephenson courtesy Rail-Online.

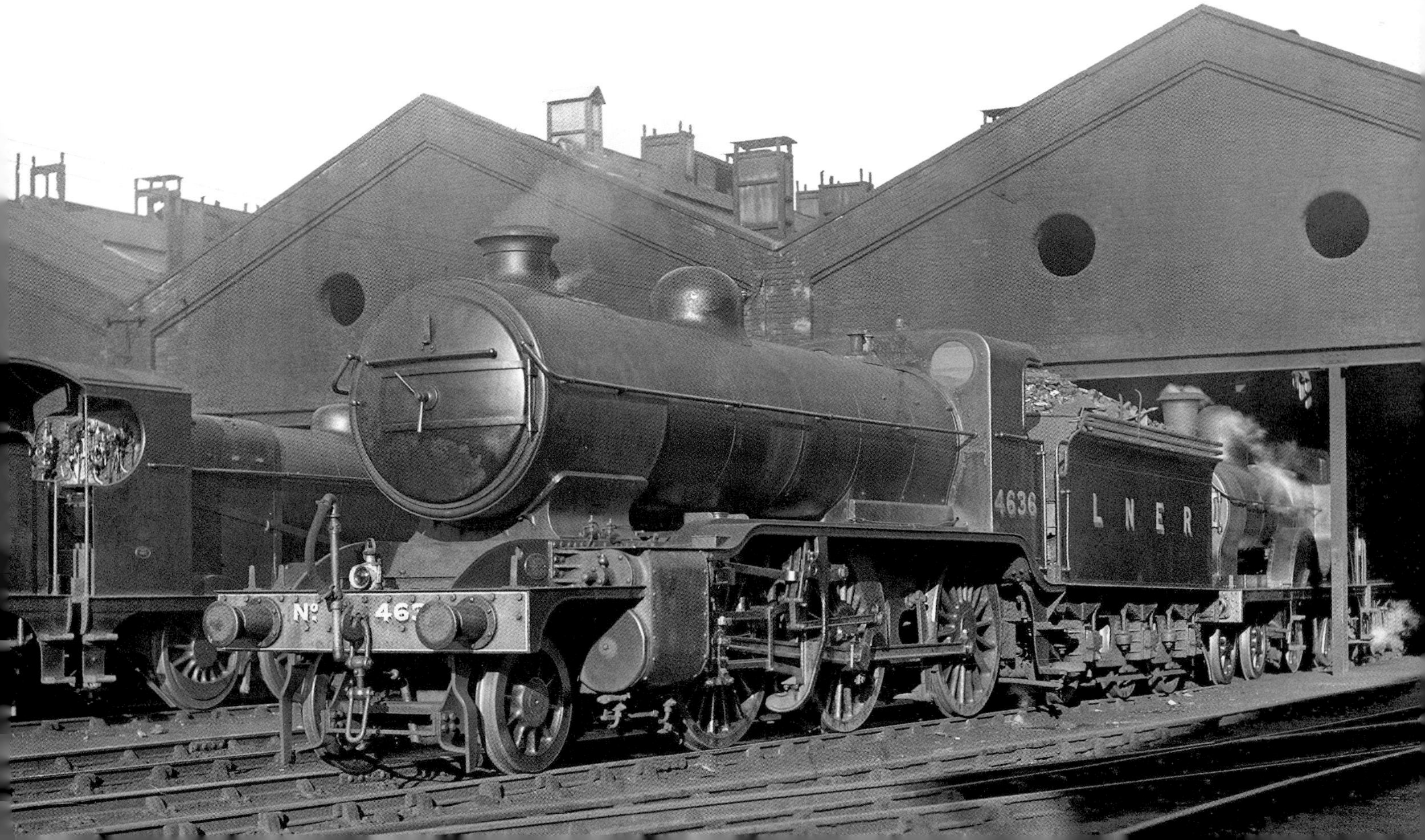

Above K2 CLASS NO. 4652

Before Grouping, the Great Eastern Railway entered into an arrangement with the Pullman Company to use their carriages on some of the continental boat services. The venture proved unsuccessful and some of the 26 carriages were redeployed, such as to the 'Harrogate Pullman'. Yet, a number of carriages, mainly restaurants, remained on the Great Eastern Section as part of other formations. One Pullman train was formed for the Sunday-only 'Clacton Pullman' which usually had seven coaches, consisting of a first-class saloon and six third-class saloons. This operated from 1922 to 1928 and in July of the latter year, K2 Class no. 4652 has been caught with the train near Romford passing Crowlands signal box. Photograph courtesy Rail-Online.

Below **K2 CLASS NO. 4650**
At the formation of the LNER, the company encountered a problem with the different brake systems in use by the constituents. Vacuum and Westinghouse air brakes were used by two of the four main companies, whilst another was transitioning away from the latter. The dilemma persisted until 1928 when vacuum brakes were favoured, though by necessity Westinghouse equipment was also used for a number of years owing to the cost of conversion. The GER was a Westinghouse system and when the K1s and K2s moved into the section the apparatus was fitted to the driver's side running plate. No. 4650 transferred into the area from King's Cross in March 1928 and was allocated to Stratford. In preparation, the Westinghouse equipment was added at Doncaster Works in the last two weeks of February. No. 4650 was at several locations until post-Nationalisation and is seen here at March shed on 4th June 1939. The locomotive reported for duty there in May 1938 and remained to June 1947. Photograph by George C. Lander courtesy Rail Photoprints.

Above **K2 CLASS NO. 4656**
Even though no. 4656 was based at King's Cross shed to August 1925, Doncaster Works altered the locomotive to conform with the GER loading gauge at a general repair completed in mid-August 1924. No. 4656 was dispatched to Stratford shed in 1925 and ran in the area to September 1927 before the Westinghouse brake equipment was installed. At this time, the engine had moved on to Cambridge. In June 1926, no. 4656 has a London-bound freight train passing through Romford station. Photograph courtesy Rail-Online.

Above **K2 CLASS NO. 4659**
A northbound freight train, consisting of a number of empty wagons, passes Sandy, near Biggleswade, with K2 Class no. 4659 on 1st May 1937. The engine was allocated to Peterborough New England shed at this time after moving from Boston in September 1936. No. 4659 subsequently returned to the latter during October 1941. Photograph by Les Hanson from the David Hanson Archive courtesy Rail-Online.

Below **K2 CLASS NO. 4667**
The Norfolk Railway reached Lowestoft via a branch on the route between Norwich and Yarmouth. This was originally promoted by the Lowestoft Railway & Harbour Co. and leased by the NR which undertook the construction work. An engine shed was established to the west of the station a year after opening in 1848 and this saw use to 1883 when replaced further west with a new four-track structure. On the depot's turntable, likely during the late 1930s is K2 Class no. 4667. The engine worked at Norwich, Yarmouth and Lowestoft in the second half of the 1930s and moved on to Colchester during early 1940. Photograph courtesy Rail Photoprints.

Below **K2 CLASS NO. 4682**
Just as newly surplus K1s and K2s found roles in the Great Eastern Section, so did engines sent to the North British Section. These class members also had to be modified to fit the loading gauge and 14 moved initially between December 1924 and August 1925. A second group of six arrived in Scotland during the early 1930s, with no. 4682 amongst this number, transferring from Peterborough New England, following a long-term allocation, in October 1931. The engine is seen around this time with an express near Ardlui, south of Crianlarich. In April 1933, no. 4682 was named *Loch Lochy* though the plates are not present here. Photograph courtesy Colour-Rail.

Above **K2 CLASS NO. 4681**
An LNER no. 2 express freight is southbound at Sandy on 1st May 1937 with K2 Class no. 4681. The company had three classifications of express freight services which had braked wagons as part of the formation. In the no. 2 category, two vehicles needed to be vacuum braked and this number increased incrementally with the number of wagons in the formation. At the minimum the cattle wagon and van are therefore braked in this formation. Built by Kitson & Co. in June 1921, the engine was a servant at Peterborough New England pre-war, with a two-year spell at King's Cross in the early 1930s and a move to Boston at the end of the decade breaking the 18-year period. Photograph by Les Hanson from the David Hanson Archive courtesy Rail-Online.

Above **K2 CLASS NO. 4684**
The North British Railway had a lineage of 4-4-0 locomotives and the final development was William Reid's K Class, introduced in 1913 with a superheated boiler. Cowlairs Works built 32 class members up to 1920. The K Class, later LNER D34 had a haulage limit of 190 tons but at Grouping this was breached, resulting in the use of two locomotives at the head of the formation. The arrival of the Gresley K Class engines was in response to this challenge, though with a capacity of 220 tons, they also required assistance on occasion usually in the form of a D34. No. 4684 *Loch Garry* has paired with an unidentified 4-4-0 to work this service which has been caught south of Rannoch. Arriving at Glasgow Eastfield during November 1931 as part of the second wave of locomotives deployed, no. 4684 was named in July 1933. Photograph courtesy Colour-Rail.

Opposite **K2 CLASS NO. 4683**
Fourteen miles north of King's Cross at Brookmans Park, K2 Class no. 4683 has a local train formed from a vintage set of carriages c. 1930. The locomotive appeared from Kitson & Co. in June 1921 and was allocated to Peterborough New England shed. In addition to the many freight duties at the depot, there were several passenger trains, particularly those stopping at all stations, such as an early morning departure from Peterborough to King's Cross, or taking over one originally from Leeds to London. The set assembled for this train features a trio of six wheelers, a twin articulated set, a twelve-wheel clerestory brake and what appear to be vans at the rear. Photograph courtesy Colour-Rail.

Above **K2 CLASS NO. 4691**
No. 4691 was amongst the original group of 14 K2 Class locomotives allocated to Glasgow Eastfield shed in the mid-1920s. During September 1932, the engine became the first of the group to be modified to have a side-window cab and more were dealt with up to 1935 when the task was complete. No. 4691 initially had a shorter roof, though was later modified to the subsequent standard. In between these two changes, no. 4691 was named *Loch Morar*. The locomotive has an express near Mallaig before the alterations were carried out. Photograph courtesy Colour-Rail.

Opposite **K2 CLASS NO. 4688**
No. 4688 was one of a pair of K2 Class locomotives loaned to York shed in August 1925 for tests on duties in the North Eastern Section. This lasted for eight months before no. 4688 returned to Doncaster. During this period two Raven B15 Class 4-6-0s were also loaned to Doncaster for comparison purposes. These events were nearly a decade behind the locomotive when pictured at Potters Bar with a suburban train. This consists of articulated sets which were introduced in the Great Northern Section for the London suburban services at Grouping. They were built up to 1930 and numbered nearly 100 examples. Photograph courtesy Colour-Rail.

Above **K2 CLASS NO. 4691**
No. 4691 *Loch Morar* is at Mallaig shed in 1936. The locomotive has the black livery with red lining whilst the number is in gold transfers with red shading. A variation existed in the early 1930s where those working in Scotland and maintained by Cowlairs Works had large cab numerals (10 in. high) whilst the tender lettering was a smaller size (7½ in.). Photograph by H.N. James courtesy Colour-Rail.

Opposite above **K2 CLASS NO. 4704**
No. 4704 *Loch Oich* was the last K2 Class locomotive built in September 1921 by Kitson & Co. At the end of 1924, the engine left Peterborough New England for Glasgow Eastfield, though two months later moved on to Edinburgh St Margaret's depot which developed a small group for use on freight trains between Glasgow and Edinburgh, to Newcastle, as well as Perth and Dundee. No. 4704 returned to Eastfield after a couple of years and remained employed to just before the war. The K2s were originally to work only goods trains on the route to Fort William and Mallaig but the class soon proved suitable for passenger duties. No. 4704 has an unfitted freight here in the late 1930s near Ardlui. Photograph courtesy Colour-Rail.

Opposite below **K2 CLASS NO. 4701**
The D34 Class was not the only option for working heavy trains along the West Highland Line as another K2 was often on hand to assist. No. 4701 *Loch Laggan* has been paired with another class member to haul this express which has paused at Rannoch station. Photograph courtesy Colour-Rail.

London & North Eastern Railway K3 Class

Above **K3 CLASS NO. 33**
In order to improve the locomotive stock in use across the newly formed LNER, Gresley produced Group Standard classes. For mixed traffic, he chose to adapt the recently introduced GNR H4 Class which became the LNER K3 Class. An initial order for 25 was placed at Darlington but soon after another 25 locomotives were requested. Production continued in relatively small batches from several sources up to 1937 when 183 had been built for the LNER. No. 33 was the fourth engine completed as part of the first order and was in service for September 1924. The locomotive is seen in works grey livery here at the time.

Opposite **K3 CLASS NO. 17**
The first K3 Class engine built under the LNER was no. 17 which appeared from the former North Eastern Railway's workshops at Darlington. Some of the design work was also carried out there and this resulted in the adoption of a side-window cab. The locomotives conformed to the new loading gauge which allowed ease of transfer between the sections and a new tender was produced which became a Group Standard design and spread to other classes. In this image, no. 17 is seen at Neasden shed in 1945. Starting life in the North Eastern Section, during the war the locomotive moved to the Great Central Section, spending nearly four years at Woodford Halse. Photograph by Colling Turner from Rail Archive Stephenson courtesy Rail-Online.

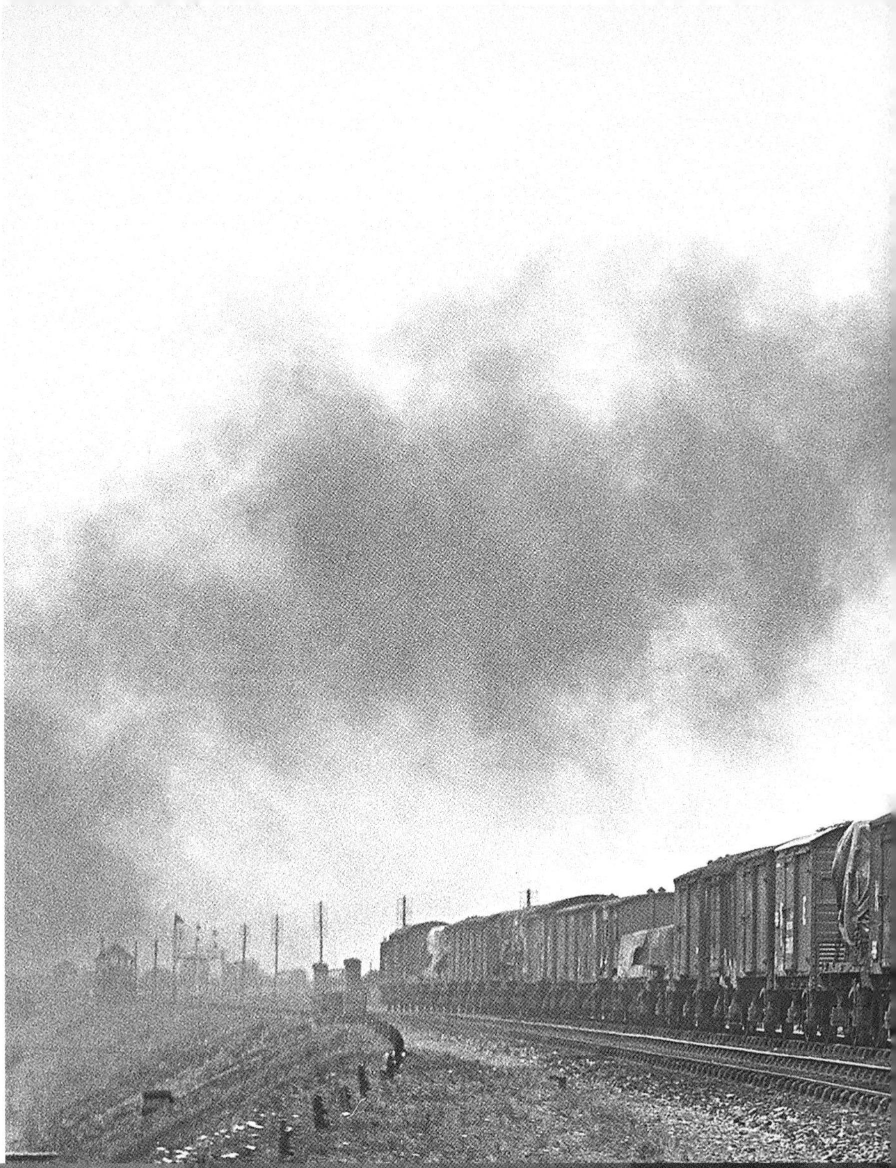

Below K3 CLASS NO. 92

When the LNER K3 Class locomotives were built at Darlington they were assigned a number which was free in the North Eastern Railway stock list. This was not continuous except in a few instances. No. 92 was completed in November 1924, following on from no. 91, though the next engine was no. 109. From this time to around 1928 the number was placed on the tender side and the works plate was on the side of the cab. No. 92 was new to Peterborough New England shed and is being coaled at the stage there on 31st July 1926. The allocation lasted for another year when a transfer to Doncaster occurred. Photograph by T.G. Hepburn from Rail Archive Stephenson courtesy Rail-Online.

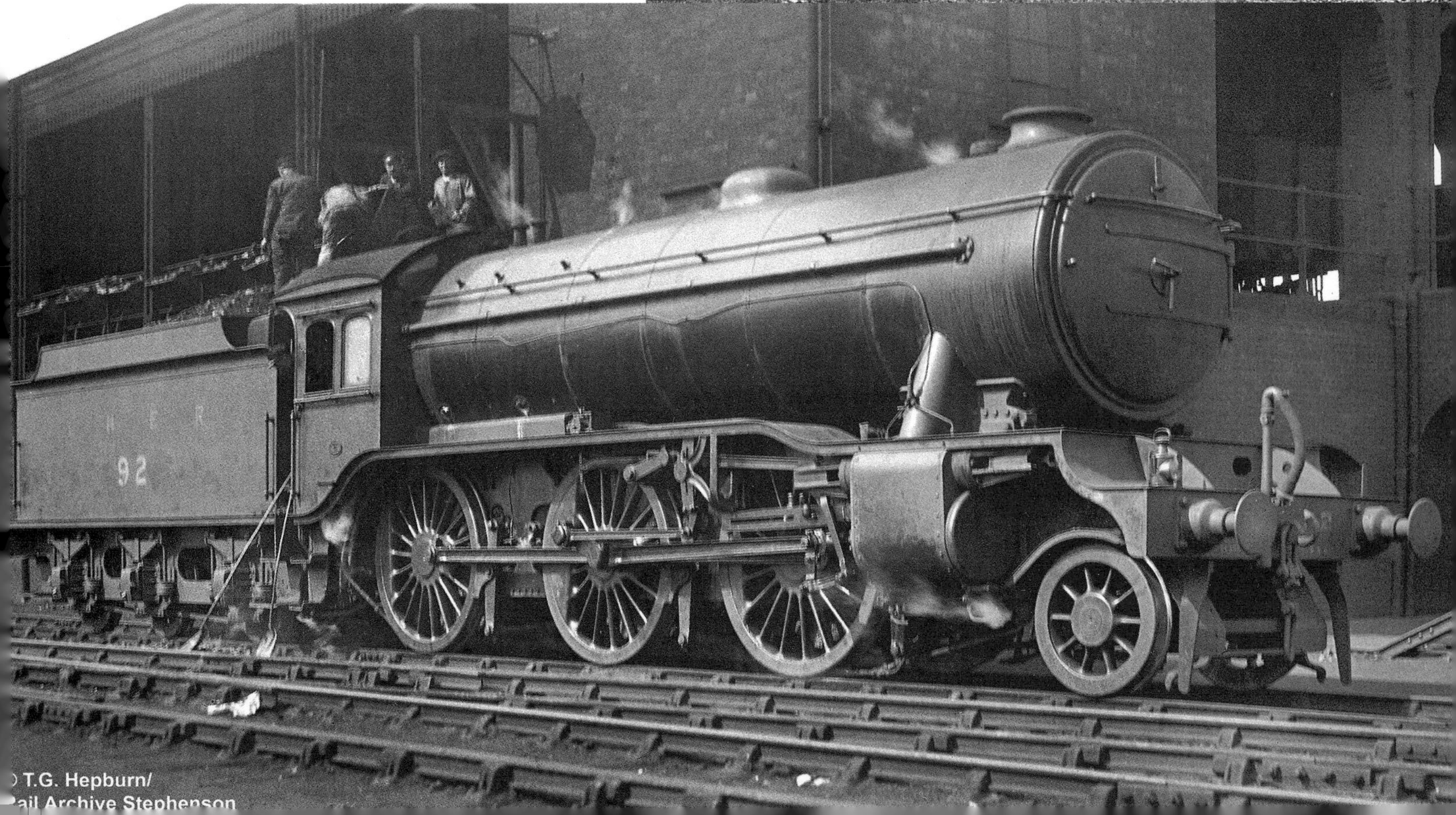

T.G. Hepburn/
ail Archive Stephenson

Above K3 CLASS NO. 80
A complicated set of circumstances resulted in the pairing of locomotive no. 80 with the GNR Type B tender which is running behind. The first two K3s built at Darlington were ready before the design of the Group Standard tender and Doncaster Works was obliged to supply two GNR Type B tenders. When later locomotives were set to be delivered to the Scottish area, the length of the turntables in use was a factor as the LNER K3 was 1 ft 1 in. longer than the GNR H4s. Two of the six were ultimately given Group Standard tenders as Doncaster only sent four, believing the original pair would be part of the exchange. The need to swap the latter two engines at work in the North Eastern Section resulted in two Great Northern Section locomotives being relieved of their Group Standard tenders, one of which was no. 80. The engine continued with the type, though briefly with a different version, until withdrawal. No. 80 is seen at Barkston junction, where the line from Nottingham to Sleaford connected with the GNR main line, north of Grantham, with a fully fitted express freight, c. 1934. Photograph by T.G. Hepburn from Rail Archive Stephenson courtesy Rail-Online.

Above **K3 CLASS NO. 113**
The longest unbroken sequence of numbers for the first group of LNER K3s was nos 111 to 114. Three of these were new to Peterborough New England, including no. 113. Before the war, the engine switched to March shed. Whilst before this time the K3s were restricted to the joint line from March to Doncaster, just as the conflict was to start the class received permission to work into the Great Eastern Section. No. 113 was at March for just over a decade and around 1946 the engine has been caught at Grantham station with a local service. Photograph by T.G. Hepburn from Rail Archive Stephenson courtesy Rail-Online.

Opposite **K3 CLASS NO. 112**
A large number of the first LNER K3 Class locomotives was allocated to Peterborough New England shed for the fast freight and other goods duties there. By the late 1930s, 36 K3s were stabled on site, though following the war class members were transferred elsewhere. No. 112 was delivered to Peterborough New England in December 1924 and had 14 years' service there. After a couple of months on the Great Eastern Section, the engine began an association with the Great Central Section, lasting to the mid-1950s. Around 1930, no. 112 is northbound away from Grantham with a fully braked express goods service. Photograph by T.G. Hepburn from Rail Archive Stephenson courtesy Rail-Online.

Above **K3 CLASS NO. 116**
Darlington carried out the design work on the Group Standard tender and based this on the North Eastern Railway's 4,125-gallon capacity type. The new tender had increased capacities, with 7½ tons of coal carried and 4,200 gallons of water. Water pick-up apparatus was provided and the scoop was located behind the middle axle diverting water into a secondary water tank via a pipe. No. 116 has the scoop down and is collecting water from Langley troughs which were located between Knebworth and Stevenage. Just 6 in. deep and 1,780 ft long, the troughs allowed locomotives to collect around 2,000 gallons of water in approx. 20 seconds. The next set of troughs was at Werrington, 50 miles northwards. No. 116 has an express from Newcastle bound for King's Cross here c. 1927. Photograph by F.R. Hebron from Rail Archive Stephenson courtesy Rail-Online.

Opposite **K3 CLASS NO. 114**
Ten miles from London, no. 114 has a local train at Hadley Wood during 1938. The local station sat immediately between two tunnels, with the northern one shorter at 232 yards, whilst the south tunnel was 384 yards. No. 114 was new to Peterborough New England in December 1924 and had a 19-year-long allocation, moving on to Mexborough depot. Photograph by C.R.L. Coles from Rail Archive Stephenson courtesy Rail-Online.

Above **K3 CLASS NO. 120**
The cab for the K3 Class followed the design of contemporary NER locomotives and had two side windows which improved the modest enclosure offered by the GNR H4s. The first 30 had the windows 3½ in. lower than those that followed as complaints were made by enginemen that they bumped their heads whilst looking out. No. 120 illustrates the problem as the driver is leaning out with cap pressed firmly against the window top. Whilst the issue was addressed, the problem was not solved and at the end of the 1920s all engines built up to that point were to have the windows moved up either 6½ in. or 3 in. depending on when they were built. This process was done as a priority when the engines passed through the workshops and took around a year to complete. No. 120 has been pictured running past Greenwood signal box, near Hadley Wood, with a northbound unfitted freight during 1925. Photograph by F.R. Hebron from Rail Archive Stephenson courtesy Rail-Online.

Opposite **K3 CLASS NO. 125**
A depressed British coal market caused the General Strike of May 1926 as colliery owners wanted to further reduce wages and increase productivity to maintain profit margins. Colliery workers carried out a strike in protest and were joined in solidarity by other unions including the railways. Some services were maintained, however, by volunteers stepping into enginemen's roles. No. 125 has such a crew on board here as the engine works the 15.00 'parliamentary' all stations train from King's Cross to Peterborough through New Southgate. The strike lasted over a week before called off with the unions unsuccessful in their aims. No. 125 was built in January 1925 and new to Ardsley shed, Leeds, but moved to Peterborough New England a month later. Photograph by F.R. Hebron from Rail Archive Stephenson courtesy Rail-Online.

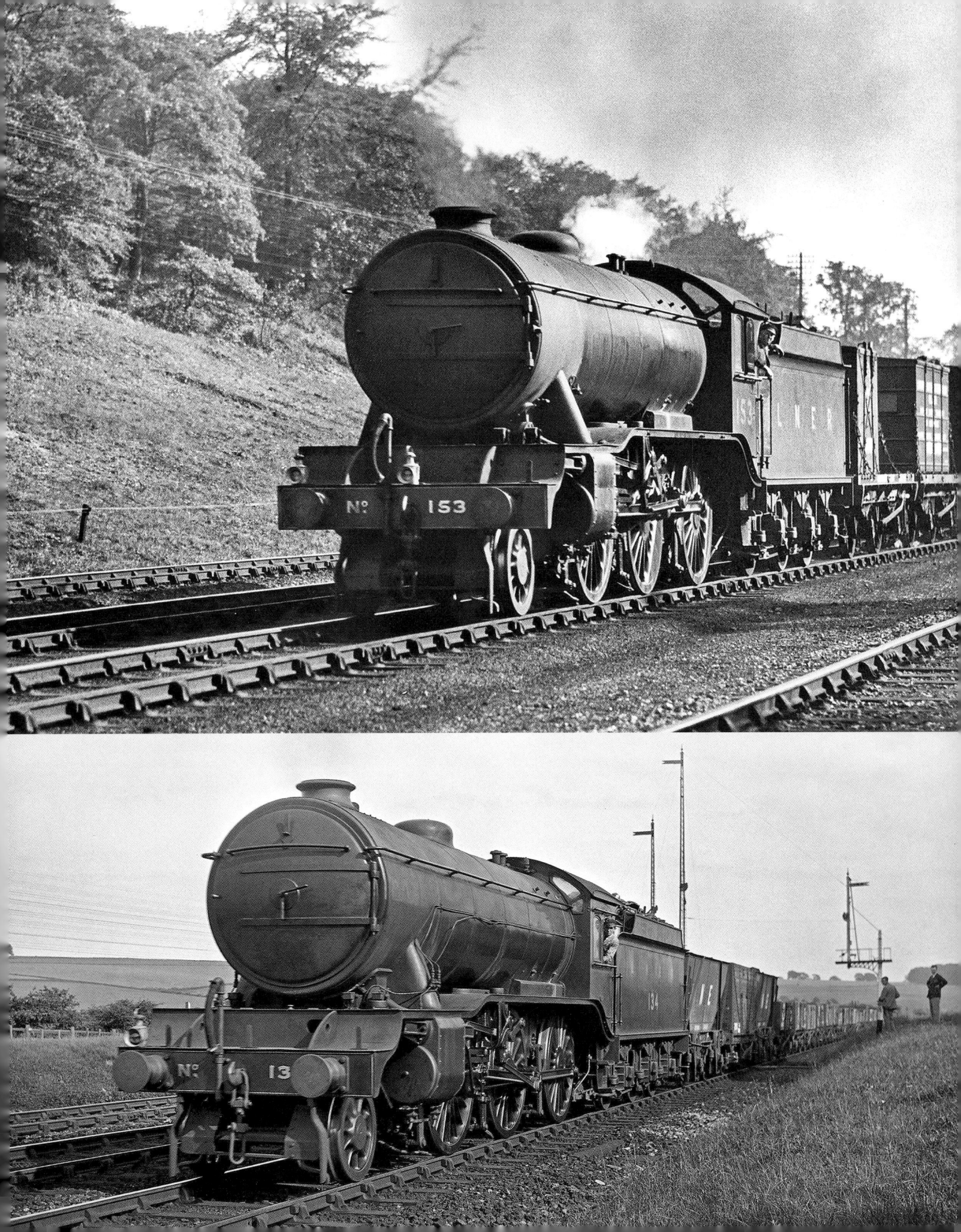

Opposite above **K3 CLASS NO. 153**
A number of Darlington-built K3s spent a short time at Ardsley shed, Leeds, when completed before moving on to permanent accommodation in the mid-1920s. No. 153 was amongst this number and dispatched to Doncaster when ready in June 1925. Three years later, the engine was at Peterborough New England, then in late 1930 a transfer to King's Cross occurred. No. 153 switched to the Great Central Section at Gorton in October 1935 and has been caught before that date on the East Coast Main Line at Greenwood (New Barnet) with a freight service. Photograph by E.R. Wethersett from Rail Archive Stephenson courtesy Rail-Online.

Opposite below **K3 CLASS NO. 134**
No. 134 is near Grantham, likely in the mid- to late 1920s. At the head of a train of empty mineral wagons, the driver appears to be hemmed in by the low cab windows. Another feature of the first group of LNER K3s was the split front spectacle window which remained the case until just before the Second World War when replaced by a single piece. No 134 was on the Great Northern Section from new to 1935. Photograph by T.G. Hepburn from Rail Archive Stephenson courtesy Rail-Online.

Below **K3 CLASS NO. 143**
Just as the LNER was improving the motive power stock in the post-Grouping period, the infrastructure also saw modernisations. At Peterborough New England shed, a project began in 1930 to install a 500-ton mechanical coaler, ash disposal plant, electric lighting and remodel the locomotive yard. The coaler is behind no. 143 here as the engine takes a rest between duties in 1934. No. 143 was not based at the depot and worked from King's Cross for the first 12 years in traffic, then moving on to Doncaster. Also visible on the left is the original coal stage which was retained to help at busy periods. Photograph by C.L. Turner from Rail Archive Stephenson courtesy Rail-Online.

Above **K3 CLASS NO. 186**
The LNER inherited a number of cattle wagons with a 10-ton capacity at Grouping and continued to add to the number in the years up to the Depression. Early examples had half doors, whilst later models had full-length doors. The traffic tended to be split into other goods formations, but in some cases whole trains of cattle were conveyed. One such service is headed by no. 186 in the mid-1930s and approaches Grantham here. The locomotive has seen the cab windows moved higher and is paired with a GNR Type B tender. The latter is present as no. 186 was a Scottish engine, based at Polmont, near Falkirk, during much of the 1930s. Doncaster Works maintained the locomotive in this decade, perhaps indicating why no. 186 is so far south and is being run-in. Photograph by T.G. Hepburn from Rail Archive Stephenson courtesy Rail-Online.

Below **K3 CLASS NO. 188**
No. 188 is lifted off the wheelset in the Crimpsall Repair Shop at Doncaster during July 1930. A Scottish engine, no. 188 is visiting over six months before the official date when Doncaster took over maintenance of the K3 Class locomotives running in the country from Cowlairs Works, Glasgow. This lasted to 1938 when the task was returned to Scotland, though Doncaster later took Darlington's responsibility for maintenance as well. No. 188 had six more visits to Doncaster during this period and when switching areas in late 1945 had a number more before condemned.

K3 CLASS NO. 229
In the late 1920s a southbound partially fitted freight approaches Potters Bar behind no. 229. Photograph by F.R. Hebron from Rail Archive Stephenson courtesy Rail-Online.

Above K3 CLASS NO. 229
Sandy was the location of a restriction in the number of running lines, meaning freight was often held to favour express passenger services. On 1st May 1937, no. 229 has stopped to allow Gresley A4 Pacific no. 2509 *Silver Link* to storm past with the southbound 'Flying Scotsman'. The two locomotives were shed mates at King's Cross during this period, though no. 229 was the senior by several years. The engine moved on to the Great Central Section in late 1939 first residing at Colwick. Photograph by Les Hanson from the David Hanson Archive courtesy Rail-Online.

Below K3 CLASS NO. 1125
No. 1125 is pictured when new from Armstrong Whitworth & Co. in late April 1931.

Above **K3 CLASS NO. 1125**
After a group of nine K3s was built at Darlington in 1930, a further 20 were ordered from Armstrong Whitworth & Co. Both batches had detail differences, with the contractor engines having steam brakes and solid hornblocks as axlebox wedges were dispensed with. No. 1125 was delivered to Doncaster shed and employed for three years before moving to Gorton, where the engine is pictured in 1934. Returning to Doncaster later in the year, the locomotive had spells at March and Peterborough New England in the mid-1930s. While at the latter, no. 1125 engaged in experiments to standardise the chimney design following the improvement of the blastpipe diameter. When this was achieved, the modification of chimney, liner and cowl was started class-wide from March 1939. Photograph from the Chris Davies Collection courtesy Rail Photoprints.

Opposite **K3 CLASS NO. 1300**
Following a gap of nearly three years, 20 K3 Class engines were ordered from Doncaster Works in 1927. The first of the group was no. 1300 which entered service in April 1929 and went to work at York. The passage of time allowed several features to be changed, such as updated cab design, improved chimney arrangement, altered valve events, screw reverser, left-hand drive, Westinghouse and vacuum brakes, drop grate, enlarged bearings, tender with flush sides, etc. No. 1300 has a freight train running southbound through Aycliffe on 26th August 1937. Photograph by C.L. Turner from Rail Archive Stephenson courtesy Rail-Online.

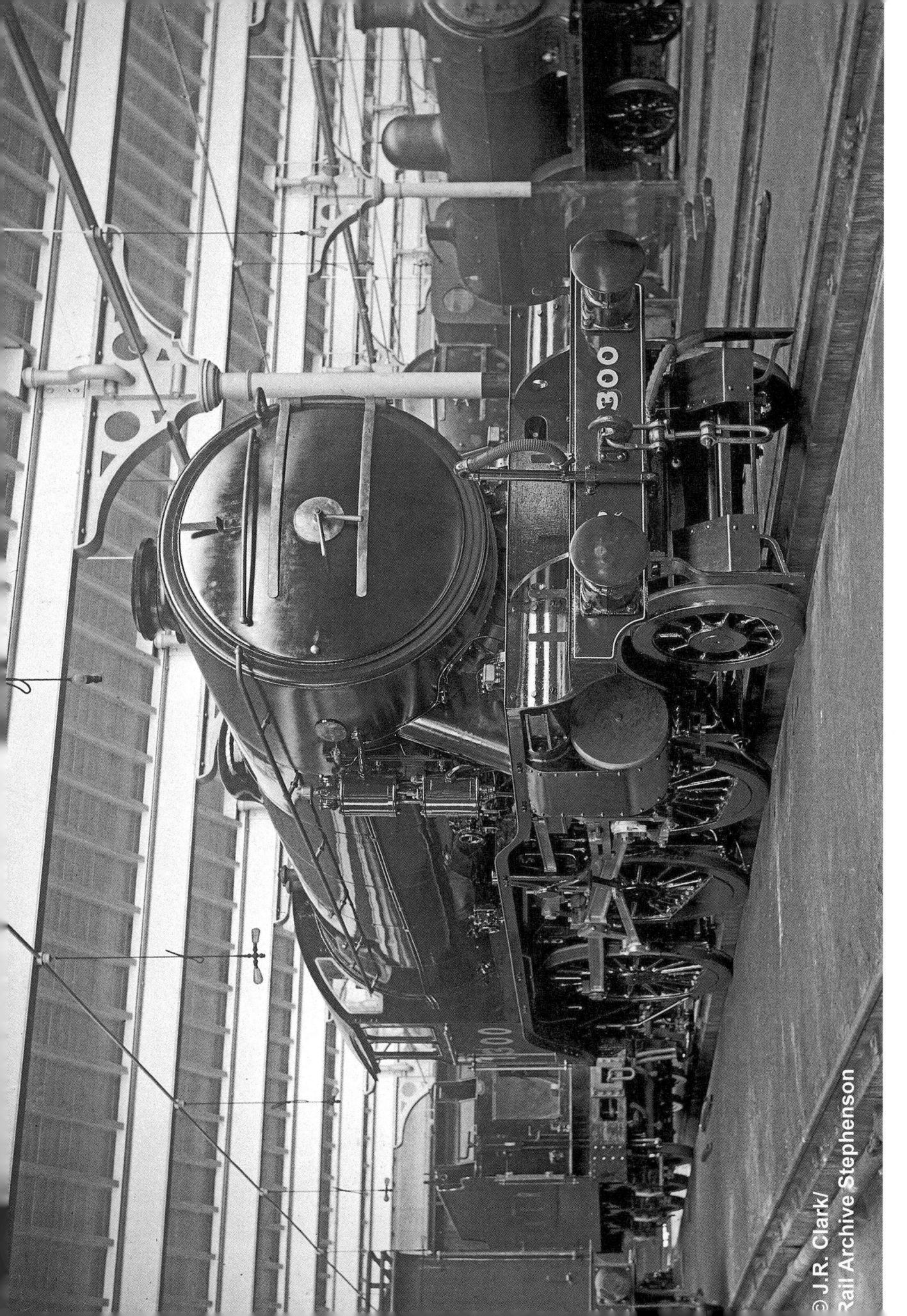

Above **K3 CLASS NO. 1397**
The 20 K3s built at Doncaster in 1929 were for the North Eastern Section and received Westinghouse brake equipment as a result. Yet, at this time steam brakes were favoured and the replacement of the Westinghouse apparatus began soon after. No. 1397 was new in December 1929 and fitted to March 1932 when steam brakes were used in tandem with vacuum. The change had only recently occurred when the locomotive was pictured on 23rd July 1932 assisting Gresley A1 Class Pacific no. 2751 *Sunstar*. The train is the 14.05 Edinburgh to King's Cross which has reached Croft Spa, south of Darlington. No. 2751 has yet to lose the Westinghouse equipment and this occurred during September 1933. Photograph by Wm Rogerson from Rail Archive Stephenson courtesy Rail-Online.

Opposite **K3 CLASS NO. 1300**
Whereas some of Gresley's earlier H2 and H3 Class had green livery, only two members of the K3 Class had the colour applied over their lifespan. Originally, H4 Class no. 1007 had green just after Grouping before the liveries were decided. Following the Second World War, an ambitious scheme to paint all locomotives, except the streamlined Pacifics, in green was quickly abandoned with just one K3 affected, no. 1935. Under the LNER the standard livery was black with red lining for the class. No. 1300 is resplendent after receiving the livery for the first time in Doncaster's Paint Shop during early 1929. Being a North East engine to 1940, Darlington would repaint the engine before Doncaster took over the task to withdrawal. Photograph by J.R. Clark from Rail Archive Stephenson courtesy Rail-Online.

Above K3 CLASS NO. 2459
Despite having a British and Empire Heavyweight boxing title fight in July 1945, contender Bruce Woodcock still had an obligation to his work as a fitter at Doncaster and is seen atop the boiler of K3 no. 2459. Born in the town, he found employment with the railways in the 1930s and in addition to his job Bruce developed a career as a boxer moving through the ranks during the 1940s. He defeated Jack London to win the title, which he held for five years, and went on to claim the European crown in the following year and this was retained until 1949. Bruce challenged for the World title in 1950 but was defeated due to a bad cut and he subsequently had to retire owing to facial injury. No. 2459 was built at the NBLC in October 1935 and worked from March shed when pictured.

Opposite above K3 CLASS NO. 2426
In January 1935, 20 K3s were ordered from the North British Locomotive Company, and these were delivered from August to November 1935, being erected at Hyde Park Works. These continued the K3 tradition of taking free numbers in the stock list between no. 2425 and no. 2468. No. 2426 was ready at the end of August and new to Doncaster shed. The locomotive has an express freight at Sandy on 1st May 1937. Photograph by Les Hanson from the David Hanson Archive courtesy Rail-Online.

Opposite below K3 CLASS NO. 2937
During February 1934, Armstrong Whitworth & Co. was contracted to build 10 K3s. No. 2397 was the final engine from the order when constructed in January 1935. Though allocated to Doncaster from new until September 1942, the engine is recorded on the Great Central Section at Wendover, south of Aylesbury on 3rd June 1939 with a local train. Photograph by George C. Lander courtesy Rail Photoprints.

Above **K3 CLASS NO. 3822**
The LNER placed two orders at Darlington Works consisting of four and 20 K3s. These appeared in 1936-1937 with no. 3822 completed in December 1936. New to York shed, around this time the depot had 14 class members employed on main line freights, as well as summer passenger services. The K3s at York were affected by the introduction of Gresley V2 Class 2-6-2 locomotives in the late 1940s, and no. 3822 departed in November 1940 for Colwick. The engine is just south of Darlington at Eryholme in the late 1930s with a northbound freight train. Photograph from the Chris Davies Collection courtesy Rail Photoprints.

Opposite **K3 CLASS NO. 4007**
No. 4007 is in the yard at King's Cross during the 1930s. A side-window cab is yet to be fitted, which was done in August 1939. The first ten H4s/K3s were altered between 1939 and 1940. Prior to this, the group followed a class-wide modification where a glass sight screen was attached for the crew to look forward with some protection. The locomotive appears to have taken a knock to the outside steam pipe cover. The interior was originally unlagged, yet this was another modification from 1934. Photograph from the David P. Williams Colour Archive.

London & North Eastern Railway K4 Class

Opposite above **K4 CLASS NO. 3441**
No. 3441 *Loch Long* has paused at Rannoch station, likely during the late 1930s. The engine was new with lined black livery in the manner of the K3 Class, yet subsequent engines had passenger green. No. 3441 was redecorated subsequently, yet later returned to the original livery as part of wartime austerity. Photograph courtesy Colour-Rail.

Opposite below **K4 CLASS NO. 3441**
In January 1937, no. 3441 *Loch Long* is seen following completion at Darlington Works in works grey livery. Note the cylinder square with lining which was a feature of Darlington-built and maintained engines.

Below **K4 CLASS NO. 3442**
As mentioned, the two-cylinder K Class locomotives struggled to work heavy traffic on the West Highland Line. In the late 1930s, the decision was made to provide a dedicated design to improve matters. Six locomotives were built between 1937 and 1939. No. 3442 *The Great Marquess* was completed at Darlington in July 1938 and dispatched to Glasgow Eastfield. Originally, the engine was christened *MacCailein Mo'r*, though was changed quickly despite the spelling apparently being correct. The locomotive is at Crianlarich station with a Fort William to Glasgow Queen Street service on 31st July 1938. Photograph by T.G. Hepburn from Rail Archive Stephenson courtesy Rail-Online.

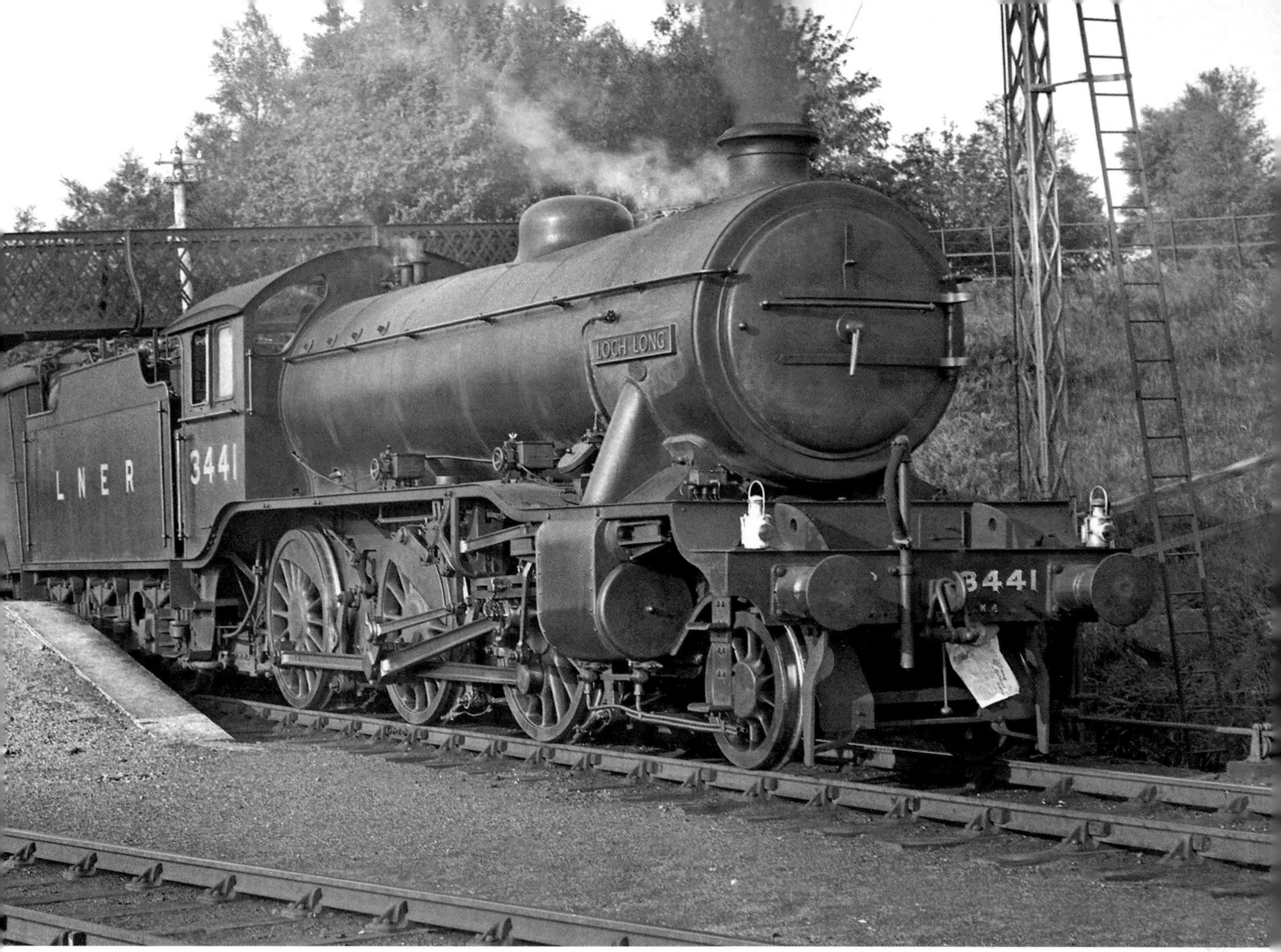
LOCH LONG
LNER
3441
3441

LOCH LONG
3441
LNER

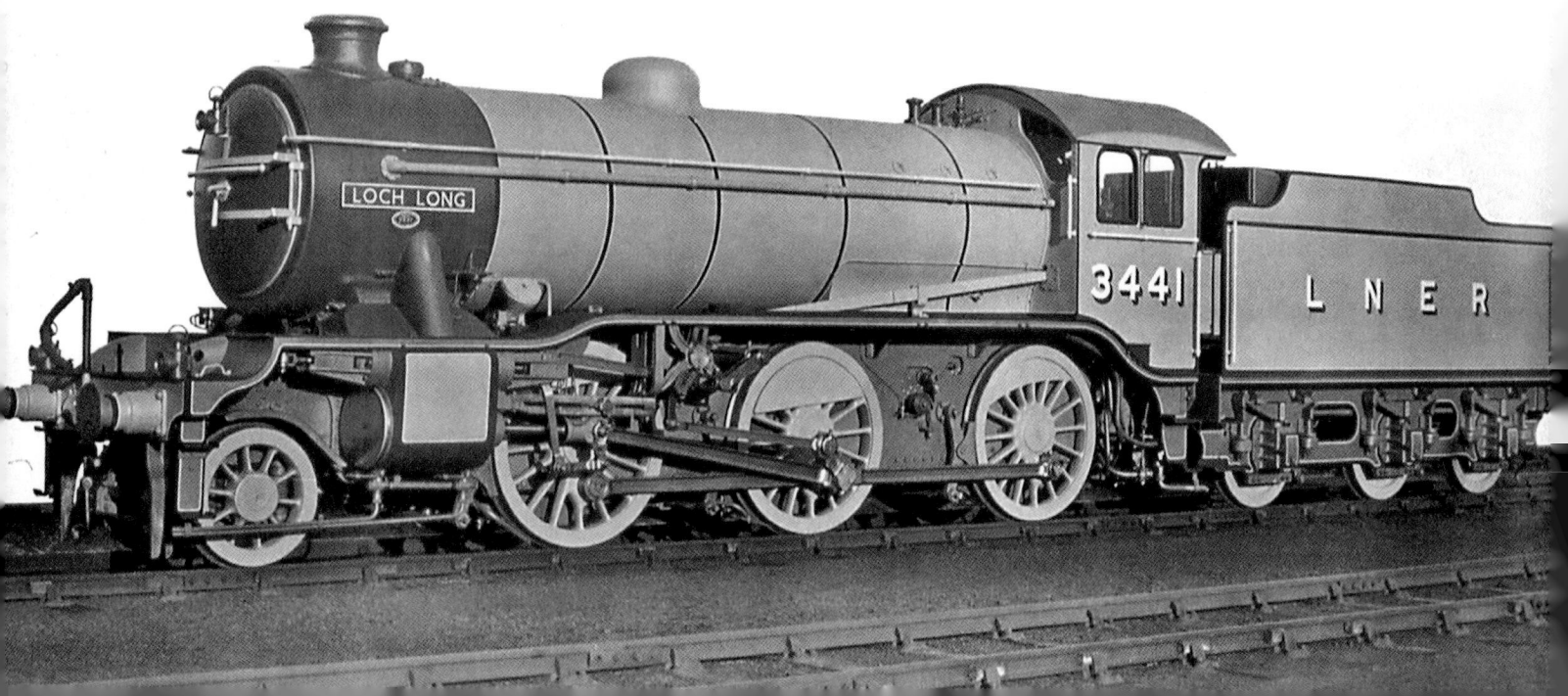

London & North Eastern Railway K2 Class 1946 Numbering

Below K2 CLASS NO. 1744
As mentioned previously, the numbering of many K3 Class locomotives was random and filled in vacant positions in stock lists. This was not just an isolated occurrence as there were numerous instances throughout the 6,263 engines in the LNER's stock. Gresley's successor, Edward Thompson, was a precise man and had a scheme produced where the whole fleet was renumbered. In this, the express locomotives started at one and a block to 999 was reserved for them and future construction. The same pattern was followed through mixed traffic locomotives to freight, then covering tank locomotives and their uses. The K Classes occupied a range of numbers from no. 1720 to 1998. K2 Class no. 4654 became no. 1744 in May 1946 as part of the renumbering, which began in January 1946 and completed a year later. The locomotive has a holiday express departing from Skegness and bound for Burton upon Trent on 13th July 1946. Photograph by T.G. Hepburn from Rail Archive Stephenson courtesy Rail-Online.

Above **K2 CLASS NO. 1790**
A pair of K2s is south of Rannoch with an express in the mid-1940s. The leading locomotive is no. 1790 *Loch Lomond*, whilst the train engine appears to be no. 1786. The latter was originally at Eastfield but spent a number of years employed by St Margaret's shed, later returning during October 1942. No. 1790 worked on the line from December 1924 to June 1954 when redeployed at Aberdeen. The locomotive received the British Railways number in April 1948. Photograph courtesy Colour-Rail.

London & North Eastern Railway K3 Class 1946 Numbering

W.H. Whitworth/
il Archive Stephenson

Above K3 CLASS NO. 1808
Original H4 no. 4008 has undergone a number of changes from GNR days. The most recent one was the switch of the number to 1808 which took place in September 1946. The side-window cab was fitted in March 1939, while a Group Standard tender has replaced the original GNR type. The locomotive also had Detroit sight feed lubrication for the cylinders yet this later caused problems and a Wakefield mechanical lubricator was substituted, seen here to the left of the steam pipe. No. 1808 has an express from Manchester London Road station to Sheffield Victoria station at Guide Bridge station around 1947. Photograph by W.H. Whitworth from Rail Archive Stephenson courtesy Rail-Online.

Opposite K2 CLASS NO. 1733
No. 1733, previously no. 4643, had carried the new number for around six weeks when pictured on 27th July 1946. The locomotive has an express on the Firsby curve which connected the Boston to Grimsby line with the branch to Skegness. Colwick-allocated, no. 1733 arrived there from Peterborough New England in early 1929. No. 1733 moved on to Immingham in April 1947 and under BR was reallocated to Scotland at Parkhead, Glasgow. Photograph by T.G. Hepburn from Rail Archive Stephenson courtesy Rail-Online.

Above **K3 CLASS NO. 1870**
A small detail changed on new K3 Class engines from 1929 was the buffer beam corners were cut to give a radiused edge. The earlier locomotives were later altered to conform. No. 1870 was the first engine new with the feature as no. 1300 in April 1929. No. 1870 from March 1946, the locomotive is travelling southwards over Wilford bridge, Nottingham, with a mid-afternoon express on 16th August 1947. The bridge crossed the River Trent a distance south of Nottingham Victoria station. The site of the latter was not viable for goods and a station was established on the other side of the bridge seen in this image, with the signal box on the left controlling movements to the facility. Photograph by T.G. Hepburn from Rail Archive Stephenson courtesy Rail-Online.

Opposite **K3 CLASS NO. 1824**
In the late 1920s, the LNER experimented with valve events and achieved a marked saving in coal consumption. This led to many classes being modified, including the K3s from the late 1920s. No. 1824, as no. 80, was modified in a general repair completed during January 1932. Later in the decade, the engine was fitted with a steam heating connection and this is visible at the front end here dangling next to the draw hook and chained to the right buffer. Another change made is seen above as the hinged hatch to access the Gresley 2 to 1 lever was a later addition owing to contamination from the smokebox increasing wear. From the mid-1930s no. 1824 was a Great Central Section engine and was based at Annesley when pictured. The locomotive is crossing Wilford bridge, Nottingham, with a train from Scarborough to Leicester on 16th August 1947. Photograph by T.G. Hepburn from Rail Archive Stephenson courtesy Rail-Online.

London & North Eastern Railway K5 Class

Above **K5 CLASS NO. 1863**
Gresley's successor Edward Thompson favoured the use of two cylinders instead of three in most applications. The K3 Class provided a donor for a prototype two-cylinder 2-6-0 in 1945. No. 1863, as no. 206, was that engine and entered Doncaster Works in February, emerging four months later. Many of the components were replaced, primarily being the cylinders, which were the same design as the B1 Class 4-6-0, and boiler. The latter was similar to the K3 type but was redesigned with thicker plates to allow a higher pressure – 225 lb per sq. in., compared to 180 lb per sq. in. – to be used, compensating for the loss of the middle cylinder. In service, no. 1863, subsequently classified K5, outperformed two K3s on trial but the ten other rebuilds ordered were cancelled. No. 1863 has been held at Greenwood signal box with a freight service during 1946. Photograph by Robert Brookman from Rail Archive Stephenson courtesy Rail-Online.

Opposite above **K2 CLASS NO. 61723**
Two members of the H2 (K1) Class were rebuilt to H3 (K2) Class specifications before Grouping whilst the remainder were dealt with between 1931 and 1937. Relatively little modification was necessary and the main changes were made to the cab. No. 61723's conversion took place over the Christmas period of 1932-1933. The engine then continued in traffic to November 1959. This date was around 18 months away when caught at Nottingham Victoria station on 21st June 1958. For the last 13 years employment was at Colwick shed and this is noted on the buffer beam, along with the class designation. Photograph by G.H. Hunt courtesy Colour-Rail.

Opposite below **K2 CLASS NO. 61739**
No. 61739 is seen at the end of a career spanning 45 years. The engine stands in the yard at Colwick around 1958 and withdrawal occurred in February 1959. Photograph by Bill Reed.

British Railways K2 Class

Above **K2 CLASS NO. 61737**
Before the Second World War seven K2 Class locomotives worked from Colwick shed and this number doubled before Nationalisation whilst 19 were allocated in the mid-1950s. No. 61737 was briefly part of the group between October 1952 and June 1955. The engine moved on to Sheffield Darnall but was condemned in November 1956. No. 61737 is in Colwick shed yard on 19th April 1955. Photograph by Bill Reed.

Opposite above **K2 CLASS NO. 61740**
On 8th July 1958, no. 61740 has been pictured on the turntable at Leicester (GC) shed. This was located to the south of Leicester Central station and opened during 1897, servicing locomotives to 1964. No. 61740 was visiting at the time of the image, travelling from Immingham. The allocation covered most of the BR period and after six months at Colwick, no. 61740 was withdrawn in January 1961 to be scrapped at Doncaster. Standing in the background is Stanier Class 5 4-6-0 no. 44945. Photograph by Tony Cooke courtesy Colour-Rail.

Opposite below **K2 CLASS NO. 61742**
Thirteen of the 75 K2 Class locomotives met their end at Doncaster Works. No. 61742 was built there in May 1916 and is starting the scrapping process in this image from mid-1962. The engine was condemned at this time whilst at Peterborough New England shed. Photograph courtesy Colour-Rail.

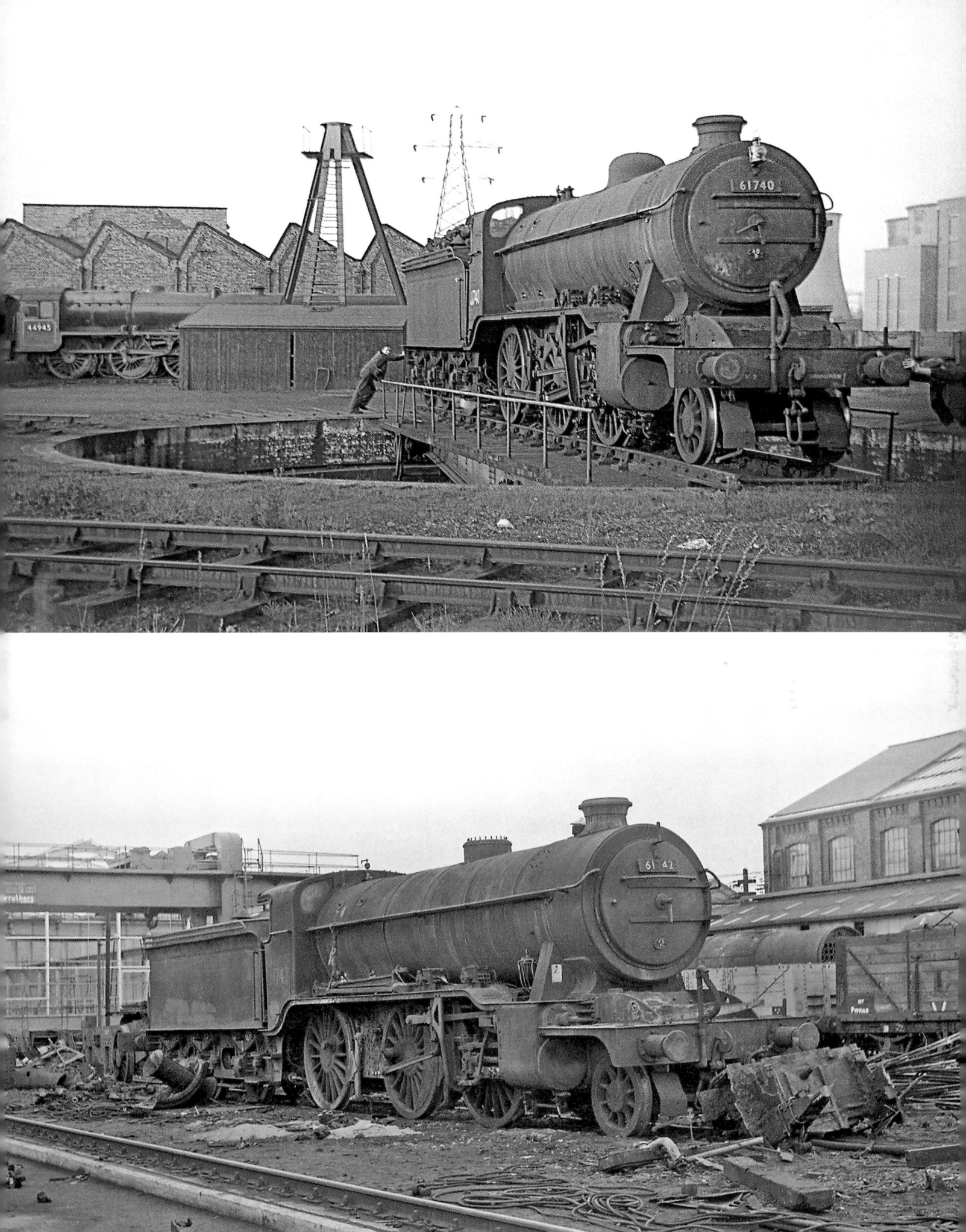

Above **K2 CLASS NO. 61749**
A total of seven K2 Class locomotives were scrapped by Darlington Works. No. 61749 has likely arrived for the task to be carried out here in the late 1950s. The last allocation was to Colwick. Photograph by Bill Reed.

Opposite above **K2 CLASS NO. 61751**
The rear of the K2 Class locomotives was slightly wider than the tender and to present a neater appearance the cab side sheets were curved inwards. This is illustrated in this image of no. 61751 which stands against the platform at Worksop station on 9th April 1955. Approaching the end of a ten-year spell at Colwick, the locomotive made a switch to Lincoln during June 1956. Photograph by Geoff Warnes.

Opposite below **K2 CLASS NO. 61745**
The first K2 was allocated to Boston in 1936 and worked a freight service between there and Doncaster, as well as some passenger trains to the coast. Before the war the engine was reallocated, but by Nationalisation there were six class members on the roster at Boston which had around 50 locomotives stabled there. The number of K2s increased subsequently and around ten were on hand for duties through to the late 1950s. No. 61745 arrived in March 1953 and was there to March 1959. During July 1957, the locomotive has been pictured at Basford, Nottingham, with a local service. When condemned in November 1960, no. 61745 was amongst the last K2s in traffic. Photograph by Bill Reed.

Above **K2 CLASS NO. 61760**

Standing outside Boston shed on 22nd October 1950 is no. 61760. At a general repair completed in April 1950, the locomotive had the new British Railways number applied after using the second LNER number for over three years. The engine had been delivered from the NBLC in June 1918 as GNR no. 1670. A new feature under BR was the front numberplate which was usually fixed on the smokebox door above the handrail and this necessitated moving the top lamp iron higher. No. 61760 was at Boston from June 1946 to July 1955 when transferred to Sheffield Darnall for the last five years of service. Photograph by T.B. Owen courtesy Colour-Rail.

Opposite above **K2 CLASS NO. 61757**

Under BR, mixed traffic locomotives received black livery with red and cream lining to engine and tender. Initially with the company name on the tender, an emblem was used, starting August 1949. A grant of arms was obtained in the late 1950s and the crest from the design was chosen to replace the emblem. Some K2s were being withdrawn at this time meaning a number never saw the change from emblem to crest. No. 61757 was part of this group as the locomotive managed to run from April 1955 to February 1959 without undergoing works attention. In February 1959 no. 61757 was condemned and sent to Darlington Works for disposal. The locomotive is pictured there awaiting the torch. Photograph by Bill Reed.

Opposite below **K2 CLASS NO. 61753**

No. 61753 appears to have experienced several problems with the smokebox as at least two patch plates are visible. Also, the fresh black paint suggests severe scorching of the bottom half of the smokebox door. In the late 1940s, the task of maintaining the K2s passed to Cowlairs and visits at other workshops became uncommon. No. 61753's last visit to Scotland occurred in mid-1955 and the engine is seen at Colwick shed in 1958-1959. The depot used the locomotive from October 1952 until condemned during September 1959. No. 61753 had to travel to Scotland for scrapping at Motherwell Machinery & Scrap, Wishaw. Photograph by Bill Reed.

Above **K2 CLASS NO. 61761**
As the 1950s progressed, the K2 Class increasingly found themselves without suitable tasks to perform. Some of those employed on the Eastern Region were sent to the ex-GCR shed at Darnall, Sheffield, initially to be placed in storage, though a group was in steam. A total of nine locomotives were allocated in 1955 and by the following year four engines were running on the lines radiating from Sheffield. No. 61761 was amongst the number and has been pictured at Sheffield Victoria station in the late 1950s. The locomotive was to become a stationary boiler in the early 1960s, being kept in running stock, at King's Cross but was evidently not suitable for the job as withdrawal occurred after a couple of weeks. Photograph from the G. Parry Collection courtesy Colour-Rail.

Opposite **K2 CLASS NO. 61763**
A train of empty stock approaches Basford North station led by no. 61763 on 7th July 1956. The station was on the ex-GNR line to Derby and Burton upon Trent from the 1870s and later a connection was made with the Great Central Railway which built the London Extension through the area in the late 19th century. There were four junctions between the two lines and no. 61763 is coming off Bulwell South curve to join the ex-GNR line. Carriage sidings were located off the ex-GCR route a short distance to the south (right of this image). No. 61763 was allocated to Colwick at the time and this covered the period August 1946 to November 1958. Photograph by T.G. Hepburn from Rail Archive Stephenson courtesy Rail-Online.

Below **K2 CLASS NO. 61765**
Engines in the Great Eastern Section received Westinghouse brake equipment in 1927/1928, amounting to 20 examples, and this was used in partnership with vacuum brakes. At the front end of the locomotive, the vacuum pipe was moved to the right side of the draw hook (as viewed from head-on) and on the left was the pipe for the air brakes. No. 61765 was equipped in early 1927, yet the full apparatus was not operational until the end of the year. The locomotive was based at Stratford for 13 years, then had spells at March and Colchester before returning in September 1947. No. 61765 is seen with a Liverpool Street to Clacton-on-Sea service at Colchester North station during the early 1950s. In the latter part of the decade, some GE Section engines had the Westinghouse components removed and no. 61765 was modified by 1955. Photograph courtesy Rail-Online.

Above K2 CLASS NO. 61766

View west towards Basford North station as no. 61766 moves away with an eastbound local passenger train on 31st July 1959. New from the NBLC in August 1918, the locomotive was prepared for the GE Section in mid-1927 and provided with Westinghouse equipment before the end of the year. The engine was first sent to Cambridge but was moved around a number of times through to the mid-1930s. During the 1940s, no. 61766 worked at March, Colchester and South Lynn. Renumbered 1766 in November 1946, at Nationalisation the locomotive had the E prefix added in February 1948 and ran to August 1949 when the BR number was applied. At Boston from March 1953, the engine had the Westinghouse brakes removed at a general repair completed in November 1954. Some time during the 1950s, no. 61766 became part of an anomalous group of K2s that had the sight feed lubrication pipes concentrated on the left-hand side, no longer running along the right-hand side of the boiler. No. 61766 was condemned at Colwick in January 1961. Photograph by T.G. Hepburn from Rail Archive Stephenson courtesy Rail-Online.

Above **K2 CLASS NO. 61767**
Out of service against a set of buffers at Colwick shed is no. 61767. The engine moved from Immingham in June 1960 and had six months at Nottingham before condemned. Photograph by Bill Reed.

Opposite above **K2 CLASS NO. 61770**
The GNR arrived in Nottingham via an independent company which built a line from Grantham. Originally sharing facilities with the Midland Railway, competition led to a new station being erected, Nottingham London Road. With the development of the Great Central Railway's London Extension, the GNR sought a connection which was achieved south of Nottingham Victoria at Weekday Cross Junction. K2 Class no. 61770 is pictured at this point, leaving the GC main line to take the spur to the original GNR route, whilst Thompson B1 Class no. 61157 runs northward with a London to Manchester express on 31st May 1949. No. 61770 was an English engine to January 1951 when dispatched for service in Scotland. Photograph by John P. Wilson from Rail Archive Stephenson courtesy Rail-Online.

Opposite below **K2 CLASS NO. 61768**
The K2 Class had two long guard irons attached to the locomotive frames at the front end for much of their career. This was in addition to a pair of guard irons coupled to the pony truck axle shield. The long guard irons were deemed redundant in the 1950s and systematically removed. No. 61768 is now without them on 19th April 1955 and the engine stands in the yard at Colwick. Settling there in August 1946, no. 61768 remained until condemned during January 1959. Photograph by Bill Reed.

Below **K2 CLASS NO. 61772**
The GNR used a top lamp iron with a decorative piece at the bottom to connect with the smokebox door. Subsequently a number of K2 Class locomotives had this replaced by an LNER Group Standard type which had a straight piece of metal to serve as the joining piece. No. 61772 *Loch Lochy* has the GNR-style lamp iron still in place here. The engine also has a smokebox door stop placed between the hinge strap that was an addition. No. 61772 is pictured at Craigendoran station on 8th August 1959. Withdrawal occurred in November 1959. Photograph by D.J. Dippie.

Above **K2 CLASS NO. 61771**
An independent scheme was promoted to connect Leicester with Newark on the GNR main line in the early 1870s. This met resistance but awakened the interest of both the GNR and London & North Western Railway for improved connections and access to mineral traffic. The two companies were able to reach an agreement for a new joint line from Market Harborough to Bottesford, as well as a connection with the Grantham to Nottingham route, with GNR-only branches to Leicester and from Newark to Bottesford. The joint line was ready throughout in the early 1880s. No. 61771 is running on the route between Melton Mowbray and the junction for Nottingham at the site of Long Clawson & Hose station on 22nd August 1959. Passenger services had been withdrawn in 1953 owing to extremely low usage and excursions were the only passenger traffic seen, such as this one from the East Coast to Leicester Belgrave Road station. The whole line was closed in 1964. Photograph by T.G. Hepburn from Rail Archive Stephenson courtesy Rail-Online.

Above **K2 CLASS NO. 61780**
Under the LNER the locomotives fitted with sight feed lubricators had these located in the cab and the oil lines ran from there along the sides of the boiler. Originally, the pipes were uncovered, yet a number of engines subsequently had metal covers fitted over them, including no. 61780. The locomotive was new from Kitson & Co. in July 1921 as no. 1690 and at Grouping was one of a small group of K2s to have the N suffix added to the number. Following Nationalisation no. 61780 was used by the engine from February 1950 and the numberplate would have been fitted at this time and is seen here below the GNR-style top lamp iron bracket. On 14th April 1958, no. 61780 has been pictured in the yard at Colwick shed, with Gresley J39 Class 0-6-0 no. 64827 on the left. No. 61780 was condemned at the depot in October 1959. Photograph by Bill Reed.

Opposite **K2 CLASS NO. 61778**
No. 61778 arrives at Sheffield Victoria station on 6th September 1958. Moving to the GE Section in 1928, the engine spent 22 years in that area before moving back to the GN Section during the 1950s. After spells at Lincoln and Boston, no. 61778's final move was to Immingham in June 1957. Withdrawal occurred in October 1959. Photograph by Geoff Warnes.

W.J. Verden Anderson/
il Archive Stephenson

ARISAIG

Above **K2 CLASS NO. 61784**
Scottish Region engines often had a modified buffer beam where holes were drilled along the bottom edge and a fitting was present in the centre below the middle lamp iron. This allowed a snow plough to be installed at quick notice. No. 61784 has the arrangement in this image from 9th May 1959 as the engine works an express at Fort William. Based at the town from May 1957, no. 61784 was also noted as being allocated to the sub-shed at Mallaig. The locomotive was withdrawn in March 1961. Photograph courtesy Colour-Rail.

Opposite above **K2 CLASS NO. 61783**
Mallaig shed opened on 1st April 1901 on the west side of the station and built by the West Highland Railway. On the turntable during June 1950 is no. 61783 *Loch Shiel*. At variance with standard practice, Cowlairs Works started painting K2 Class locomotives with green livery from August 1947, including a pair of English engines. Soon confined to Scottish engines, the livery continued to be applied until July 1948. No. 61783 was a late recipient in June 1948 and remained in that condition to January 1955, though during that period, the engine was paired with a tender painted black. In this image, 'British Railways' is on the tender and the Cowlairs practice of having a larger number and smaller lettering continues. Photograph by W.J. Verden Anderson from Rail Archive Stephenson courtesy Rail-Online.

Opposite below **K2 CLASS NO. 61786**
A passenger train has made a stop at Arisaig station, to the south of Mallaig, with no. 61786 in charge of the service. The engine was at Eastfield for two years until 1927 when becoming one of the seven class members at St Margaret's depot. The latter lost the allocations by Nationalisation and no. 61786 returned to Eastfield by the end of 1942. The locomotive was condemned in December 1959. Photograph courtesy Rail-Online.

Above **K2 CLASS NO. 61791**

No. 61791 *Loch Laggan* has a through freight bound for Mallaig at Lochailort on 1st June 1950. Built at Kitson & Co. in August 1921, the engine left Peterborough New England for Glasgow in March 1925. Briefly at St Margaret's early in the war, no. 61791 was associated with the West Highland Line to withdrawal in March 1960. The locomotive was named in May 1933 and had a cab installed during 1934. Photograph by V.J. Verden Anderson from Rail Archive Stephenson courtesy Rail-Online.

Opposite **K2 CLASS NO. 61787**

The West of Scotland came under the attention of railway promoters in the 1840s, yet a viable scheme was not developed until the early 1880s. This was produced by private individuals and the North British Railway but met much resistance and the Bill was turned down by Parliament. Interest in social issues in the area ultimately forced through the West Highland Railway from Craigendoran to Fort William in 1889. Construction started soon after and took around five years, with the formal opening taking place on 11th August 1894. Fort William station was built close to the site of the ferry crossing with three platforms. One of the lines extended a short distance past the station to reach the jetty for shipping. No. 61787 *Loch Quoich* stands on this track during August 1959, with the partially castellated station building off to the right. As part of road improvements, the line to the dock was removed in the early 1970s and the station was also demolished to be replaced by a modern structure. Photograph by R. Trethewey courtesy Colour-Rail.

Above **K2 CLASS NO. 61788**

Over a four-year period, GNR tenders were built with a handrail following the line of the cab cut out whilst the side sheet curved sharply down. From 1913, this gap was removed and the tender side sheet extended the length of the top. As the tender was a standard type, the versions became intermixed over the years and despite being a late example, no. 61788 *Loch Rannoch* has a pre-1913 tender here, even retaining the GNR-type lamp irons. A change made in the 1950s saw the water pick-up apparatus removed from the K2 Class in both England and Scotland, with those in the latter country being modified from 1955 and no. 61788 was dealt with during 1956. The engine is seen outside Glasgow Eastfield shed on 14th August 1960 and survived almost another year before withdrawal. Photograph by K.C.H. Fairey courtesy Colour-Rail.

Opposite **K2 CLASS NO. 61787**

Just as the K2 Class was drafted to the West Highland Line in the 1920s, British Railways Standard Class 5 4-6-0s arrived at Glasgow Eastfield in the 1950s to assist on services. No. 73077 was new there in May 1955 and employed to 1962 when displaced by diesel locomotives. The engine is taking the lead here at the head of a Glasgow train which is leaving Crianlarich in the late 1950s. No. 61787 *Loch Quoich* is connected to the train. Photograph from the David P. Williams Colour Archive; original monochrome photograph by Donald Luscombe.

Above **K2 CLASS NO. 61791**
No. 61791 *Loch Laggan* appears to be undergoing maintenance at Fort William shed on 31st July 1958. The engine also looks to have been the victim of a rough shunt, with the front running plate next to the cylinder fractured. Around two months earlier, the locomotive had been in Cowlairs Works for attention, but the next visit was for disposal in April 1960. Photograph courtesy Colour-Rail.

Opposite **K2 CLASS NO. 61793**
As the K2s were displaced after Nationalisation, new areas of work had to be found. In the mid-1950s, a small group assembled at Aberdeen Kittybrewster shed and these were used on both passenger and freight services to places such as Boat of Garten, Elgin, Keith, Macduff and Inverness. No. 61793 was allocated in September 1952 but moved out to Keith during July 1954 for similar duties. In February 1959, no. 61793 was condemned and disposed of at Inverurie Works, a short distance to the north west of Aberdeen. Photograph by P.J. Hughes courtesy Colour-Rail.

V.J. Verden Anderson/
I Archive Stephenson

British Railways K3 Class

Above **K3 CLASS NO. E1839**
A total of 25 K3 Class locomotives received the 'E' prefix at Nationalisation, though there were several variations, including some with the company branding on the tender. The first saw the letter placed before the number, then moving above. British Railways lettering was on the tender in 7 in. high lettering and subsequently changed to 8 in. high. No. E1839 was amongst the group and has the second variants in both instances in this image of the locomotive at Princes Risborough with a Niddrie to Marylebone goods train during March 1948. Photograph by H.K. Harman from Rail Archive Stephenson courtesy Rail-Online.

Opposite above **K2 CLASS NO. 61792**
No. 61792 was another K2 which was moved on from Glasgow Eastfield shed to Aberdeen Kittybrewster in the early 1950s. Like no. 61793, the locomotive was there for a couple of years before moving on to the ex-Great North of Scotland Railway system at Keith where four class members gathered initially before joined by another two, all of which survived to 1959-1960. No. 61792 has just left Boat of Garten with the 08.30 freight service to Keith during September 1957. Photograph by W.J. Verden Anderson from Rail Archive Stephenson courtesy Rail-Online.

Opposite below **K2 CLASS NO. 61794**
To ease maintenance, engines stabled at Glasgow Eastfield had drop grates fitted during the mid-1930s. This was operated by a rod running from the fireman's side of the cab (left-hand side) to the firebox and was visible above the running plate. No. 61794 *Loch Oich* was based at the shed for the modification to be carried out during February 1936. The locomotive was named in mid-1933 and had the cab fitted at a general repair late in 1934. No. 61794 has an express at Glasgow Queen Street High Level station on 15th August 1959. The locomotive had another year in service left. Photograph by D.J. Dippie.

K3 CLASS NO. 61801
Near Chelmsford, no. 61801 has an eastbound train of cattle wagons on 3rd October 1959. Photograph by K.L. Cook from Rail Archive Stephenson courtesy Rail-Online.

Above K3 CLASS NO. 61800
Brocklesby station was a short distance west of Grimsby and Immingham. No. 61800 passes through light engine on 18th June 1961. Doncaster-allocated at this time, the locomotive had another year left in service. Photograph courtesy Rail-Online.

Below K3 CLASS NO. 61810
No. 61810 has been diverted into sidings at High Dyke, south of Grantham, for the southbound 'Flying Scotsman' to pass on 16th September 1961. Photograph by Hugh Ballantyne courtesy Rail Photoprints.

Opposite K3 CLASS NO. 61813

The North Eastern Railway preferred to use steam-powered reversing apparatus on the company's locomotives making enginemen familiar with this method. In 1925 no. 141 was fitted for steam reverse and proving satisfactory, other class members in the North Eastern Section were modified, including no. 61813 during late 1927. The design of the mechanism was later altered and most engines were changed. Yet, incidents were recorded where overtravel of the middle valve caused damage, which was a known problem, though thought to be accentuated by the steam reverse. The GN Section engines had the equipment removed soon after but the NE Section lasted to the mid-1930s, with no. 61813 affected in December 1934. The engine passes through Hessle station with a freight train in the 1950s. Photograph courtesy Rail-Online.

Below K3 CLASS NO. 61816

From late 1925 new K3s built had left-hand drive and earlier class members were altered subsequently. No. 61816 was built at Darlington in September 1924 as no. 39 with right-hand drive. The change occurred by November 1934 and at the same time the valve events were improved. Later changes include a switch to a flush-sided Group Standard tender, higher cab windows, provision of footsteps at the front end. No. 61816 is pictured at Colwick shed on 12th July 1952. The allocation covered the period August 1950 and September 1953. Photograph by Bill Reed.

BRITISH RAILWAYS K3 CLASS

K3 CLASS NO. 61819
No. 61819 has a freight train at Doncaster station during 1954. The engine was a long-term Hull resident. Photograph courtesy Colour-Rail.

Above **K3 CLASS NO. 61821**
Mapperley tunnel on the GNR's Nottingham to Derby line suffered from mining subsidence and had to close in 1960. No. 61821 was engaged to work on the last day and has the 12.35 from Nottingham Victoria running tender first on 2nd April. Photograph by T.G. Hepburn from Rail Archive Stephenson courtesy Rail-Online.

Below **K3 CLASS NO. 61821**
A long-term Colwick resident, no. 61821 has a local train travelling towards Nottingham Victoria at Bagthorpe Junction on 20th September 1958. Photograph by Bill Reed.

Above **K3 CLASS NO. 61824**
Rothley station opened with the Great Central Railway's London Extension on 15th March 1899 and was located between Leicester and Loughborough. No. 61824 arrives with the 17.15 Leicester Central to Nottingham Victoria station on 23rd May 1959. Rothley survived another four years but was later resurrected as part of the Great Central Railway heritage line. Photograph by Tony Cooke courtesy Colour-Rail.

Opposite above **K3 CLASS NO. 61829**
No. 61829 crosses the River Witham a short distance to the west of Lincoln station with a northbound local train during the 1950s. Under BR, the locomotive was allocated to Stratford from February 1949 to February 1957 when transferred to Peterborough New England. This stay lasted to September 1962 when moved on to Doncaster, though withdrawal occurred soon after and disposal was carried out at the town's works at the end of the year. Photograph by T.G. Hepburn from Rail Archive Stephenson courtesy Rail-Online.

Opposite below **K3 CLASS NO. 61832**
Though late in the day for steam, Doncaster Works still repaired no. 61832 and returned the engine to traffic during June 1961. The locomotive is seen stripped in the Crimpsall Repair Shop during the process on 14th May. The K3s were repaired at the main workshops responsible for each area of the LNER and later British Railways' Regions, with visits of necessity made to other facilities at times. No. 61832 ran until October 1962 and the last three years passed at Woodford Halse. Photograph courtesy Rail-Online.

61829

61832

Above **K3 CLASS NO. 61861**
No. 61861 has a local train at Norwich on 2nd August 1959. The engine had completed a final general repair just a few weeks earlier and still retains the sheen of the repaint. The livery was British Railways' standard mixed traffic livery of black with red lining and the tender crest is the second version. No. 61861 was one of the engines that had the 'E' prefix at Nationalisation, this being used from March 1948 to May 1949. When pictured, the engine was allocated to March shed. Photograph by D.L. Dott courtesy Colour-Rail.

Opposite above **K3 CLASS NO. 61836**
The K3s were first allocated to Immingham in 1940 as four locomotives arrived for duties, mainly on ex-GNR lines in Lincolnshire. Seven more joined the group in 1943 and amongst these was no. 61836, as no. 125. The allocation lasted to February 1959 and in that time, Immingham had increased the number to around 15. No. 61836 has an express freight running southward on the ex-GCR main line at Bagthorpe Junction, Nottingham, on 11th June 1957. Photograph by Bill Reed.

Opposite below **K3 CLASS NO. 61833**
Though some of the earlier right-hand drive K3s were converted to left-hand drive in the 1930s, others remained as built throughout their careers. No. 61833 was erected at Darlington as no. 118 in December 1924 and remained with right-hand drive to the end in September 1961. The locomotive stands outside Colwick shed on 15th May 1958. Photograph by Bill Reed.

K3 CLASS NO. 61864
A mineral train passes through Welwyn North station with no. 61864 on 15th August 1962. Photograph by D.J.Dippie.

Above **K3 CLASS NO. 61868**
Collecting a train from Portobello goods sidings, Sunderland, is no. 61868. Seen on 8th September 1960, the engine was March-allocated and had just over a year left in traffic, though much of this time was spent at Colwick following a transfer. Photograph by D.J. Dippie.

Below **K3 CLASS NO. 61870**
A Skegness to King's Norton, Birmingham, express passes Ancaster on 3rd August 1957 with no. 61870. Photograph by T.G. Hepburn from Rail Archive Stephenson courtesy Rail-Online.

Above **K3 CLASS NO. 61883**
No. 61883 was new from Doncaster in October 1929 and destined for the North Eastern Section where the engine remained until condemned during December 1962. With Westinghouse brakes at this time, the apparatus was removed just two years later. Following the Second World War, a large number of K3s were sent to Hull Dairycoates for traffic to and from the docks and no. 61883 reported for duties in October 1947. Many remained there throughout the BR period, including no. 61883. In the late 1950s, the locomotive has been caught at Cellars Clough, Marsden, west of Huddersfield, with a York to Ardwick freight service. Photograph by Kenneth Field from Rail Archive Stephenson courtesy Rail-Online.

Below **K3 CLASS NO. 61880**
The Royston & Hitchin Railway was completed in October 1850 and already had plans to connect with the Eastern Counties Railway at Shepreth, with the latter company providing a short branch to join with the line to Cambridge. The GNR initially declined to work the R&HR which was leased to the ECR for a short period. When this expired, the GNR proposed to take over operations and this was accepted by the Great Eastern Railway, with a platform provided at Cambridge station and a separate goods shed erected to the north west. The building stands in the background here on 12th August 1959, with the GNR branding still visible on the gable end 36 years after the company's demise. No. 61880 of the local shed is shunting in the yard. Photograph courtesy Colour-Rail.

Above **K3 CLASS NO. 61889**
No. 61889 had around 18 months of employment at Colwick shed from October 1960 and is seen in the midst of the spell during May 1961 at Nottingham Victoria station. Photograph by Bill Reed.

Opposite **K3 CLASS NO. 61885**
After the Second World War the Scottish K3s were concentrated at Edinburgh St Margaret's shed for goods services on all radiating lines. No. 61885 left March shed for Scotland in October 1941 and was allocated to St Margaret's, though had around a year at Carlisle mid-war before returning for the remainder of the engine's career, which ended during November 1959. The locomotive is on the Waverley route and climbing from Newcastle to Whitrope summit with a Carlisle to Edinburgh freight in 1952. Photograph by E.E. Smith from Rail Archive Stephenson courtesy Rail-Online.

Above **K3 CLASS NO. 61896**
Just west of Basford North station, Nottingham, no. 61896 has a partially fitted freight, c. 1960. The locomotive was one of nine completed at Darlington in 1930. Three of these were destined for the North British Section and a small detail difference was the fitting of larger brake blocks for the tender wheels. Another change was the omission of water scoops which was also a feature of a number of other engines sent new to the NB Section. This may have been deemed short-sighted subsequently as no. 61896 left Carlisle for March shed briefly in 1938 and departed again for Sheffield permanently in 1940. From June 1954 to May 1962 when withdrawn, no. 61896 was based at Colwick. Photograph by Bill Reed.

Opposite **K3 CLASS NO. 61891**
At the west end of Lincoln station, no. 61891 halts the progress of a group of pedestrians as the engine moves off with a local train on 28th August 1954. The locomotive was working from Immingham at this time and the allocation dated from November 1940, though no. 61891 had three months at Lincoln over the summer of 1945. The engine returned to the city for two months before sent for scrapping at Doncaster in September 1961. Photograph by T.G. Hepburn from Rail Archive Stephenson courtesy Rail-Online.

Above **K3 CLASS NO. 61897**
Kielder Viaduct stands soundly nearly 100 years after completion as no. 61897 passes over with a passenger service during 1953, whilst the footbridge in the foreground twists precariously. The viaduct was built as the Border Counties Railway connected the Newcastle & Carlisle Railway with the Waverley Route. Completed in 1862 using stone, the baronial style was chosen to please a local nobleman and was also built skewed which was unusual. Traffic ran on the route to 1958 but the viaduct was later saved and forms part of a walking route. Photograph courtesy Colour-Rail.

Opposite above **K3 CLASS NO. 61910**
No. 61910 has an express freight on 15th September 1960. The engine is south of Immingham on the line to Grimsby at Roxton level crossing, with the crossing gate keeper's house visible on the right. To the left of the locomotive is Roxton Sidings signal box which appears to have controlled a minor local siding. In February 1960, no. 61910 left Gorton after a decade and joined the eight or so K3s allocated to Woodford Halse. Withdrawal occurred in July 1962. Photograph courtesy Rail-Online.

Opposite below **K3 CLASS NO. 61908**
As dieselisation progressed in the East of England, useful steam locomotives were moved to other areas awaiting new employment. No. 61908 was made redundant at Norwich in January 1960 and at first moved on to March before being placed in store at Staveley Barrow Hill, near Chesterfield. A number of engines found work elsewhere, though no. 61908 appears to have been placed in service for the depot and is seen on 7th August 1960 at Bagthorpe Junction, Nottingham, with a northbound local passenger train. The locomotive left for Mexborough in June 1961 and had another six months in service before condemned. Photograph by Bill Reed.

Above K3 CLASS NO. 61914
No. 61914 was erected by Armstrong Whitworth & Co. in May 1931 as no. 1156. The engine became no. 1914 in January 1946 and two years later had the 'E' prefix applied. The BR identification was in use from February 1949. On 9th August 1962 – a rainy summer's day – the locomotive is pictured at Uttoxeter station with a local freight train. No. 61914 approaches from the west and is likely Nottingham-bound, being Colwick-allocated at the time. The GNR's Derbyshire and Staffordshire Extension reached Burton upon Trent and met the North Staffordshire Railway which gave running powers through Uttoxeter. To the west of the town the GNR purchased the failed Stafford & Uttoxeter Railway to reach Stafford in 1881. Photograph by Bill Reed.

Opposite K3 CLASS NO. 61914
Daybrook station opened on the GNR's Derbyshire and Staffordshire Extension from 1st February 1876. The facility was initially known as Bestwood & Arnold and went through several variations before Daybrook was settled on in August 1876. Daybrook later provided a junction for the Nottingham Suburban Railway which reduced the distance to Nottingham London Road station when opened in late 1889. No. 61914 has not used this line, rather the original route via Gedling and had originated at Nottingham Victoria, using the spur from the GCR main line to the GNR route. The destination is Basford North on 15th March 1958. Photograph by Bill Reed.

Above **K3 CLASS NO. 61919**

A train of mineral wagons passes through Guide Bridge station with no. 61919 in the early 1950s. The engine was allocated to Gorton from January 1950 to June 1954 when a transfer to Lincoln occurred. No. 61919 stayed there to June 1961 when the locomotive left traffic. Photograph by Raymond McCarthy from the Eddie Johnson Archive courtesy Rail-Online.

Opposite **K3 CLASS NO. 61941**

Two locomotives have completed their duties and are now coupled together to travel from Selby station to the local depot for servicing during the mid-1950s. Gresley D49 Class 4-4-0 no. 62731 *Selkirkshire* is in the background, whilst the focus of the image is no. 61941. The engine was one of the contingent at Hull Dairycoates shed and worked there from July 1945 to July 1961 when sent to Doncaster for scrapping. Photograph from the David P. Williams Colour Archive.

Above **K3 CLASS NO. 61944**

Gresley's three-cylinder locomotives experienced issues with the centre cylinder driving rod big end bearing overheating. For the K3 Class the bearing surface was 8¼ in. diameter and originally 5½ in. long, but from 1929 the length was increased by ½ in. The locomotives built previously remained with the original dimensions throughout their careers. As extra protection from overheating, a device filled with fluid was installed which emitted a pungent odour when a certain temperature was reached. This became a feature from mid-1938 and no. 61944 was fitted at an out of course works visit to Gorton in July 1938. The engine has a partially fitted freight at Grimsby in the early 1960s. With '40B' on the smokebox door, the locomotive was allocated to Immingham, which covered the period December 1960 to January 1962. Photograph by D. Preston courtesy Colour-Rail.

Opposite **K3 CLASS NO. 61954**

In the 1920s and 1930s, the technique of welding metal was applied to the construction of locomotives, carriages and wagons. The North British Locomotive Company kept up with the latest developments and the tenders completed for K3 Class locomotives in the late 1920s/1930s featured welded water tanks, whereas those built at Darlington continued to favour rivets in fabrication. No. 61954 was new from the NBLC in October 1935 with a welded tender. The engine has been pictured at Quay Road level crossing, Bridlington during the late 1950s. Photograph courtesy Colour-Rail.

Below **K3 CLASS NO. 61964**
A Class E freight passes through Rossington station, just south of Doncaster, in the late 1950s. The locomotive was new to the town's shed from Armstrong Whitworth & Co. in June 1936, though left three months later for King's Cross. For the next ten years, the engine was on the Great Central Section at several locations, then moving to Lincoln. Doncaster again had use of no. 61964 from November 1957 and was on the roster to the end in July 1961.

Above **K3 CLASS NO. 61973**
The K3s on the Great Eastern Section left for other areas of the system during the Second World War, though March shed had a sizable allocation meaning their presence in East England continued. Following the conflict, the class was reintroduced to the area at Norwich and Lowestoft, which received 11 and four engines respectively. No. 61973 was initially dispatched to Norwich but moved over to Lowestoft from October 1948 and remained to January 1959. The locomotives there were mainly used on fish and freight trains to London and back whilst also working local passenger trains. No. 61973 has an express passenger train here at Brentwood, between London and Chelmsford, on 18th July 1952. Photograph by D. Preston courtesy Colour-Rail.

Above **K3 CLASS NO. 61977**

In the 18th century Skegness began to develop as a resort for those in poor health and had money to travel to take the fresh sea air. The opening of the East Lincolnshire Railway from Boston to Grimsby brought Skegness closer to people, yet a rail connection was not made until the early 1870s. This prompted the Earl of Scarborough to develop his landholdings in Skegness into a holiday resort, firstly for people locally in the Midlands (Nottingham/Derby/etc.), then further afield. In the 1880s, 200,000 visited Skegness and just before the First World War around 750,000 travelled to the seaside. The town continued to be popular in the 1950s and 1960s and no. 61977 has an express ready to leave Skegness station for Nottingham and Derby, c. 1960. Photograph courtesy Rail-Online.

Below **K3 CLASS NO. 61974**

As part of the London Extension, the Great Central Railway established a large depot at Annesley to the north west of Nottingham. During the early 1930s, the first K3 Class locomotives were sent there and started on the express coal trains which ran from Annesley sidings to Woodford Halse twice daily returns, later four, which became well-known as the 'windcutter' service. The K3s were only involved for around five years before moved away. But they returned at the start of the Second World War and reached a peak number of 16 as the build-up to D-Day progressed owing to troop and munitions movement from the ex-GCR main line to the Great Western Railway system. After the conflict, the amount of K3s at Annesley reduced and was eliminated altogether when the ex-GCR lines passed to the London Midland Region in 1958. No. 61974 was at Annesley for 12 years from February 1943, then moved over to Colwick in January 1955 to December 1960 when a transfer to Immingham occurred. During the late 1950s, the locomotive has a local train at Daybrook station, Nottingham. Photograph by Bill Reed.

Below **K3 CLASS NO. 61988**

Tweedmouth was the location of a marshalling yard up to 1939 when this closed. As a result, a number of locomotives became redundant including 15 of the 17 K3s at Tweedmouth shed. Based there from 1929, the K3s worked freight services northward to Edinburgh and southward into Tyneside. The class members were redistributed to these areas, with the majority at Heaton and York and seven departed for St Margaret's shed. No. 61988 was amongst this group and had been sent new to Tweedmouth from Darlington in January 1937, leaving during March 1939. Several more K3s joined no. 61988, when LNER no. 3828, during the war and in the post-Nationalisation period around 20 were employed. In the 1939-1945 period some K3s were loaned to Edinburgh Haymarket for wartime traffic, later being returned to St Margaret's. An unexplained aspect of K3s at the two sheds and Carlisle is that many spent over 200 days out of service – as recorded in *Yeadon's Register of LNER Locomotives* – during 1944, including no. 61988. The others were: no. 61879, no. 61881, no. 61855, no. 61900, no. 61909, no. 61911, no. 61924, no. 61931, no. 61955, no. 61968, no. 61992. No. 61988 has a train of empty coaching stock running through Princes Street Gardens, Edinburgh, on 6th September 1955. The engine was one of the few K3s with a low number of allocations and withdrawal from St Margaret's occurred in November 1959. Photograph courtesy Rail-Online.

Above **K3 CLASS NO. 61978**
March-allocated no. 61978 has a Peterborough train at Norwich Thorpe station on 17th July 1955. In November of the following year, the locomotive transferred to Peterborough New England shed and was employed to August 1961 when condemned. Photograph courtesy Rail-Online.

British Railways K4 Class

Above **K4 CLASS NO. 61993**
No. 61993 *Loch Long* appears to have left the turntable line at the rear on 16th September 1957. The location is Glasgow Eastfield shed where the engine was allocated from new in January 1937 to April 1959. At the latter date a move to Thornton Junction occurred and the locomotive was active until October 1961. Photograph by M.J. Reade courtesy Colour-Rail.

Opposite above **K4 CLASS NO. 61994**
With rope attached to the handrail, some maintenance work appears to be under way at Glasgow Eastfield shed on 21st May 1961. No. 61994 *The Great Marquess* is the locomotive receiving the attention. The cylinders of the K4s followed the design of those used by the K3s and the first K4, no. 3441, also conformed in having a right-angle joint for the steam pipes connecting the smokebox and cylinders. The remaining K4s had a straight pipe as the free passage of steam was important to efficiency. No. 61994 left Glasgow Eastfield for Thornton Junction in December 1959 and when condemned during December 1961 was preserved. Photograph courtesy Colour-Rail.

Opposite below **K4 CLASS NO. 61995**
Following the Second World War, no. 61995 *Cameron of Lochiel* was the second engine to have the green livery restored in November 1947. Subsequently, the K4 Class was downgraded to the British Railways mixed traffic black livery with lining and this was applied from 1952. No. 61995 was a recipient in June of that year. A feature of some Scottish engines was the use of blue background to numberplate and nameplate which is the case for no. 61995. The locomotive is at Fort William in the late 1950s. Photograph courtesy Colour-Rail.

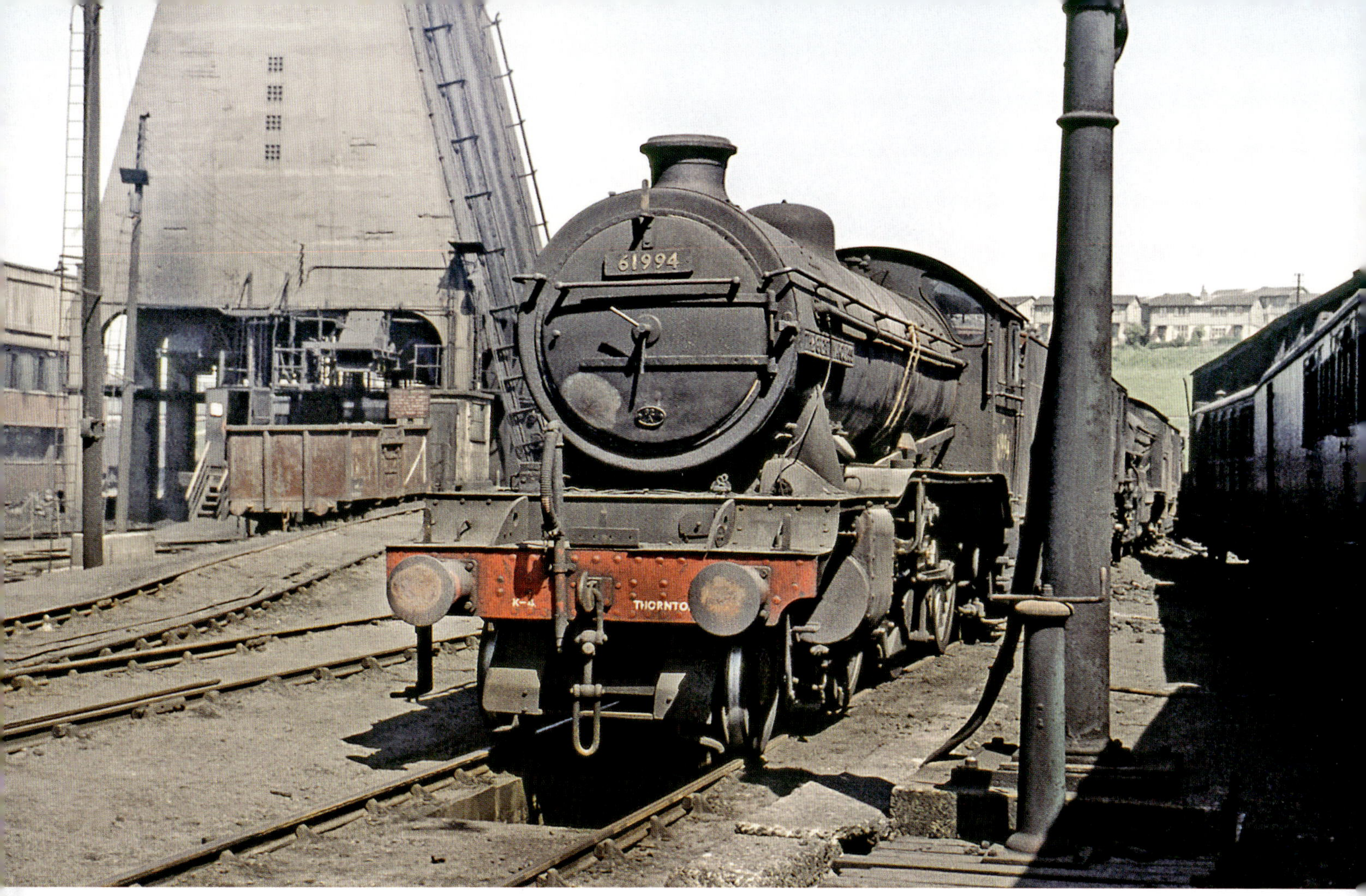
61994
K-4

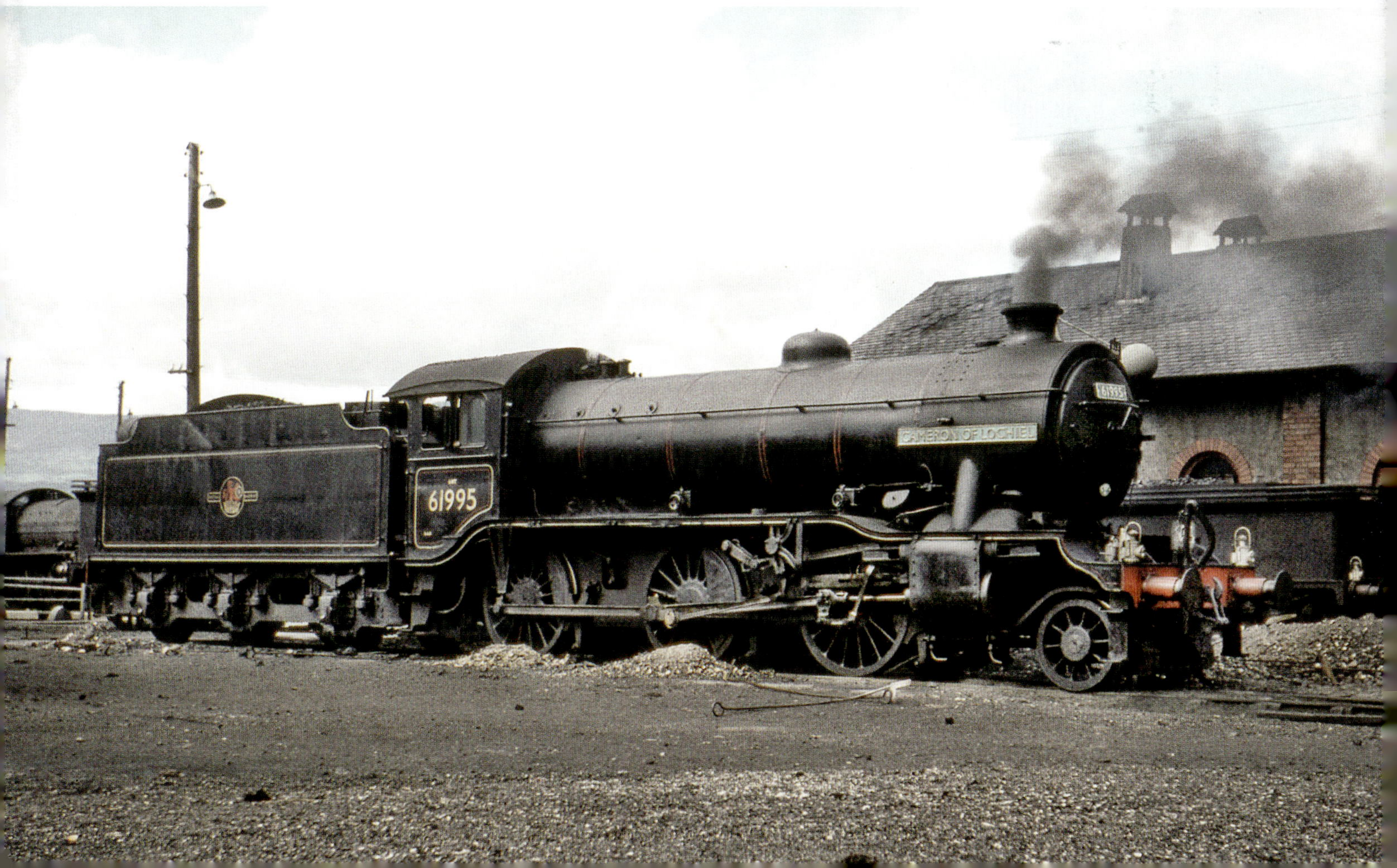
CAMERON OF LOCHIEL
61995

Above **K4 CLASS NO. 61998**

No. 61998 entered traffic in December 1938 as *Lord of Dunvegan*. There are two branches to this clan and whilst one is based at Dunvegan Castle, the clan chief is known as Macleod of Macleod. The holder of the position at the time, Dame Flora Macleod of Macleod petitioned the LNER to change the name of the locomotive to *Macleod of Macleod* which occurred in March 1939. The locomotive was repainted after Nationalisation in April 1948 to green livery, though had the British Railways number and branding on the tender. The latter is just visible under the grime here as the engine takes water at Crianlarich station while working a local train in the early 1950s. The locomotive was repainted in mixed traffic livery in February 1953. Photograph by T.G. Hepburn from Rail Archive Stephenson courtesy Rail-Online.

Opposite **K4 CLASS NO. 61996**

The West Highland Line featured a number of speed restrictions and as a result the K4s were fitted with speed indicating apparatus. This was acquired from Flaman and no. 3441 originally had this attached to the left-hand side trailing crank pin before moved to the right side and later engines were equipped similarly. No. 61996 *Lord of the Isles* was new in December 1938 and had the Flaman recorder from this time, the triangular bracket for which is visible here at the cab end. During the war, the apparatus was removed and not restored subsequently though the bracket was left in place. No. 61996 is pictured at Thornton Junction station, east of Dunfermline, during May 1959 with a local train. Photograph by G.H. Hunt courtesy Colour-Rail.

Thompson K1/1 Class

Above **K1/1 CLASS NO. 61997**
No. 61997's sojourn in England ended in November 1949 when returning to Glasgow Eastfield. The engine later moved up to Fort William during May 1954 and remained there to withdrawal in June 1961. The locomotive is between Lochailort and Glenfinnan with the 12.30 Mallaig to Fort William service in mid-1955. Photograph by W.J. Verden Anderson from Rail Archive Stephenson courtesy Rail-Online.

Opposite above **K1/1 CLASS NO. 3445**
No. 3445 *MacCailin Mór* is seen at Doncaster Works following reconstruction in December 1945. As part of Thompson's standardisation plans, the K4s were to be rebuilt with two cylinders and new boiler with higher working pressure. In the event no. 3445 was the only engine altered and served as the basis for A.H. Peppercorn's K1 Class 2-6-0. No. 3445 has unlined black livery here with shaded number transfers and abbreviated 'NE' on the tender.

Opposite below **K1/1 CLASS NO. 1997**
No. 3445 *MacCailin Mór* was renumbered in December 1946 to no. 1997. At this time the engine was allocated to Norwich and following rebuilding had moved around several times. The locomotive was first at Peterborough New England for trials, then transferred to Blaydon for two months and finally returned to Scotland. No. 3445 did not settle and had left the country by the end of 1946. During the engine's spell in East Anglia, this image of no. 1997 at Colchester shed was captured. The locomotive has the original smokebox door with widely spaced yet short hinge straps, opening knob on the left and no handrail on the door. The top lamp iron also looks to be the old GNR type. There were several different smokebox doors used subsequently with differences. Photograph by A.W. Croughton from Rail Archive Stephenson courtesy Rail-Online.

Peppercorn K1 Class

Above **K1 CLASS NO. 62001**
Though ordered in July 1947, the first Peppercorn K1 did not appear until May 1949. This was no. 62001 which was delivered to Darlington shed from the NBLC's Queen's Park Works. The locomotive was not important enough to displace D3 Class 4-4-0 no. 62000 to take the number because the latter engine was specially employed as the motive power for Officials' specials. No. 62001 is receiving coal from the plant at Darlington shed on 26th June 1964. Photograph by David P. Williams.

Opposite **K1 CLASS NO. 62003**
Like the other K Classes, Peppercorn's K1 Class qualified for British Railway's mixed traffic livery of black with red and cream lining. No. 62003's paintwork has become quite worn from the last application with significant weathering to the smokebox. The engine is in the yard, possibly out of service with the oxidisation of the wheels, at Darlington shed in the 1960s. Photograph by Bill Reed.

Above K1 CLASS NO. 62004
No. 62004 has a mineral train at Kirk Sandall, north east of Doncaster, on the line to Humberside in the early 1950s. The engine was new to Darlington and remained employed there for much of the BR period. Photograph by Geoff Warnes.

Below K1 CLASS NO. 62007
A northbound train of hopper wagons passes through Billingham station on 12th April 1966. No. 62007 of Tyne Dock shed is the locomotive. Photograph by Geoff Warnes.

Above **K1 CLASS NO. 62006**
Disturbing a track gang working at Alnmouth station during the mid-1960s is no. 62006. Photograph courtesy Rail-Online.

Below **K1 CLASS NO. 62008**
In June 1965 no. 62008 is light engine at Darlington. The locomotive was new to the local shed in June 1949 and only moved on in March 1966 to West Hartlepool. Withdrawal occurred at the end of the year. Photograph by Bill Reed.

Opposite above **K1 CLASS NO. 62017**
The Blyth & Tyne Railway's line between Bedlington and Morpeth crossed the River Wansbeck by a 400-yard-long wooden viaduct which was erected in the late 1850s. By the 1920s, the structure was in need of replacement and a steel trestle viaduct with 14 spans was constructed and stood 86 ft above the river. No. 62017 crosses Wansbeck Viaduct with a coal train for Cambois Power Station on 1st June 1966. Photograph by Hugh Ballantyne courtesy Rail Photoprints.

Opposite below **K1 CLASS NO. 62005**
No. 62005 had been involved with a railtour a few days before this image was captured at Tyne Dock shed on 13th September 1967 perhaps giving a reason why the locomotive is in a presentable condition close to the end of steam. The turntable is not similarly well maintained as trouble looks to be given by the vacuum motor and enginemen have resorted to the old-fashioned method of manhandling the apparatus to turn the engine. No. 62005 left service at the end of the year and was purchased privately for preservation. Photograph courtesy Rail Photoprints.

Below **K1 CLASS NO. 62016**
Nos 62011 to 62020 were sent to Gorton when new, yet the group was soon redeployed to March shed where another 20 K1s joined the ranks. These engines replaced some older designs running in the East of England and found use on passenger and freight duties to the principal places in the area. No. 62016 was at March from May 1950 to December 1960 and is in the shed's yard towards the end of this period. Photograph by Bill Reed.

Above **K1 CLASS NO. 62022**
A feature of new locomotives in the mid- to late 1940s was the fitting of electric lighting equipment to display headcodes at night. A Stone's electric generator was located on the right-hand side of the smokebox and produced power via a steam connection in the smokebox. Some locomotives subsequently lost the apparatus, though most retained the generator to withdrawal. No. 62022 is still fitted on 24th May 1961, with the engine in Blaydon depot's yard. Shed-mate Raven Q6 Class 0-8-0 no. 63431 is also present. No. 62022 was allocated from new in August 1949 to May 1962. Photograph by D.J. Dippie.

Opposite above **K1 CLASS NO. 62031**
When new, the K1s were delivered to Glasgow Eastfield shed for acceptance from the NBLC. The depot utilised class members on their own duties whilst this process was carried out and when the construction was finished the depot did not have another K1 until 1952 when receiving an allocation. No. 62031 was originally dispatched to March shed but moved to Scotland in February 1952, then to Fort William in May 1954. The engine is at Fort William exchanging the line token during the 1950s. No. 62031 was condemned there in December 1962. Photograph courtesy Rail-Online.

Opposite below **K1 CLASS NO. 62048**
Work-worn no. 62048 passes under St James' Bridge, Doncaster, c. 1960, with a freight train. The locomotive was new in October 1949 to Darlington and had several moves around the North Eastern Region until withdrawn in June 1967.

Above **K1 CLASS NO. 62052**
No. 62052 was new to March shed in November 1949 and later part of the original group of five sent to Glasgow Eastfield for employment on the West Highland Line. These engines were subsequently moved on to Fort William from which they operated northward and southward. No. 62052 is at Mallaig station with a passenger train c. 1960. All K1s on the route were withdrawn by the end of 1962 and no. 62052 was later scrapped at Cowlairs. Photograph by Bill Reed.

Below **K1 CLASS NO. 62037**
As K1s became surplus to requirements at the original depots, new locations received class members. From March depot, the ex-GCR shed at Retford obtained five in late 1961, followed by four more in 1962. These were briefly employed on freight and coal trains to Lincoln and Immingham, as well as local colliery traffic. No. 62037 transferred in October 1961 and left for Doncaster in November 1962. The locomotive has an excursion train at Retford station on 29th July 1962. Photograph by Geoff Warnes.

K1 CLASS NO. 62062
No. 62062 is at York station on 3rd June 1965. The locomotive was based at the city from August 1963 to August 1966. Photograph by Geoff Warnes.

Above K1 CLASS NO. 62064
Awaiting disposal at Darlington shed on 15th November 1965 is no. 62064. The engine was condemned two months earlier and the location of scrapping is unknown. Gresley A4 Class Pacific no. 60010 *Dominion of Canada* is behind and would go on to be preserved in that country. Photograph by D.J. Dippie.

Below K1 CLASS NO. 62070
No. 62070 was the last K1 to be completed in March 1950 and for some reason this was over a month after no. 62069. New to March shed, no. 62070's association with the depot lasted to June 1962 when amongst the second group sent to Retford. The locomotive remained there to January 1965. No. 62070 is at March, c. 1960. Photograph by Bill Reed.